GEOBOTANY II

GEOBOTANY II

Edited by

Robert C. Romans

Bowling Green State University
Bowling Green, Ohio

PLENUM PRESS · NEW YORK AND LONDON

Library of Congress Cataloging in Publication Data

Geobotany Conference (2nd : 1980 : Bowling Green State University)
    Geobotany II.

    ''Proceedings of the Geobotany Conference held March 1, 1980, at Bowling
Green State University, Bowling Green, Ohio.''
    Bibliography: p.
    Includes index.
    1. Paleobotany—North America—Congresses. 2. Paleoecology—North Amer-
ica—Congresses. 3. Botany—North America—Congresses. 4. Phytogeography—
Congresses. I. Romans, Robert C. II. Title.
QE935.G46   1980   561   81-13992
ISBN 0-306-40832-5   AACR2

Proceedings of the Geobotany Conference held March 1, 1980,
at Bowling Green State University, Bowling Green, Ohio

© 1981 Plenum Press, New York
A Division of Plenum Publishing Corporation
233 Spring Street, New York, N. Y. 10013

Printed in the United States of America

PREFACE

The papers and abstracts in this volume were presented at the Geobotany Conference held at Bowling Green State University, Bowling Green, Ohio, on 1 March 1980. Geobotany was again the unifying theme of the Conference.

The conference was a continuation of a series of geobotany meetings initiated more than a decade ago by Dr. Jane Forsyth of the Department of Geology at Bowling Green. Such meetings allow for interaction among geologists and botanists, palynologists and paleobotanists, ecologists and paleoecologists. It is through such interaction that a true interdisciplinary approach to geo-botanical problems can be realized.

The purpose of the Bowling Green meeting was to provide a forum for that interaction. Formal presentation of research results and informal discussions of research problems were utilized in an attempt to solve geobotanical problems.

Academic or administrative units of Bowling Green State University who sponsored the conference include the Department of Biological Sciences, the Department of Geology, the College of Arts and Sciences, and the Graduate College. The success of the meeting was due, in part, to the combined efforts of those units.

Special thanks are due to Jane Trumbull and Patricia McCann for their help in preparing the manuscripts for publication.

Robert C. Romans

Bowling Green
1981

CONTENTS

ABSTRACTS OF OTHER PAPERS PRESENTED

PRAIRIE FENS AND BOG FENS IN OHIO: FLORISTIC SIMILARITIES, DIFFERENCES, AND GEOGRAPHICAL AFFINITIES

Ronald L. Stuckey[1] and Guy L. Denny[2]

Ohio State University[1] and Ohio Dept. of Natural Resources[2], Columbus, Ohio[1,2]

Fens are alkaline wetlands occurring within the glaciated region of North America. They are characterized by an association of pioneer, now relict plants that typically inhabit sites adjacent to, on, or at the base of somewhat porous, calcareous gravel of esker-kame complexes associated with major end moraines. Here artesian springs supply clear, cold, oxygen-deficient ground water that is moderately hard containing bicarbonates of calcium, magnesium, and rarely sulfates. Upon contact with the atmosphere at the earth's surface, these compounds are precipitated forming a grayish-white lime-rich substance, or marl. This marl, combined with an accumulation of organic remains over time, becomes the primary substrate for the specific plants that colonize these habitats. The pH of the substrate is usually between 8 and 8.5. Although many species of vascular plants are good indicators of fens, some of the best examples are Potentilla fruticosa (shrubby cinquefoil), Parnassia glauca (grass-of-parnassus), Lobelia kalmii (Kalm's lobelia), Gentiana procera (fringed gentian), and Solidago ohioensis (Ohio goldenrod). In this paper, the fens of Ohio are discussed from the viewpoints of distribution in the state, successional stages in the fen meadow plant community, history of floristic studies, floristic similarities and differences, and geographical affinities and origin of the flora.

DISTRIBUTION

In Ohio, fens occur primarily in the west-central part in Champaign, Clark, Greene, Logan, and Miami counties, and in the northeastern portion in Holmes, Portage, Stark, and Summit counties. Elsewhere, they are more scattered, such as in Erie,

Seneca, and Williams counties in northern Ohio, and Fairfield,
Pickaway, and Ross counties in south-central Ohio. The distribution of 52 known fens in Ohio is shown on the accompanying map
(Figure 1), with closed symbols (dots and squares) locating 40
extant fens and open symbols (circles and squares) representing 12
fens that are believed to have been destroyed. Many of the extant
fens still contain a good representation of the species, whereas
further exploration may reveal remnants of those fens believed to
have been destroyed. Most of these latter fens are in the popu-
lated areas of the state in the vicinity of Canton, Columbus,
Dayton, and Springfield. The fens in Ohio are located totally
within the area covered by Wisconsinan glaciation, and within that
region, they are associated with esker-kame complexes near major
end moraines in the west-central and northeastern part of the
state, although the latter point is not illustrated in Figure 1.
However, when plotted on the Glacial Map of Ohio (Goldthwait,
White, and Forsyth, 1961), the distribution of fens in association
with esker-kame complexes is strikingly evident. Some of the fens
are located near the headwaters of certain major river systems,
such as the Little Miami, Big Miami, and Mad River systems in
west-central Ohio, and the Cuyahoga, Mohican, and Tuscarawas
River systems in northeastern Ohio. Fens are absent from a large
area of western and northwestern Ohio where the maximum relief
is less than 100 feet.

SUCCESSIONAL STAGES OF THE FEN MEADOW PLANT COMMUNITY

The fen meadow plant community is characterized by three
successional stages or zones of development, (1) the open marl
zone, (2) the sedge-meadow zone, and (3) the shrub-meadow zone.
In size, the open marl zone may be a few acres, but more often it
is only a few square feet or rarely absent. The zone may be
somewhat centrally located or dispersed throughout the fen. It
is characterized by shallow pools situated upon mucky well-exposed
marl or by spring-fed streamlets traversing those zones that are
successionally more mature. The plants are scattered and are
comprised primarily of low-growing sedges, rushes, grasses, and
other small herbaceous individuals, such as Triglochin maritimum,
T. palustre (arrow-grass), Tofieldia glutinosa (false asphodel),
Lobelia kalmii (Kalm's lobelia), Utricularia cornuta (bladderwort),
and Calopogon tuberosus (C. pulchellus) (grass-pink orchid). In
the spring and early summer when the marl is continuously wet, and
if vegetation is sparse, one can easily sink several inches into
the substrate while walking upon this zone.

As time progresses, the open marl zone is replaced by a
sedge-meadow zone. The remains of sedges and other plants from
the open zone begin to accumulate and a thin deposit of sedge

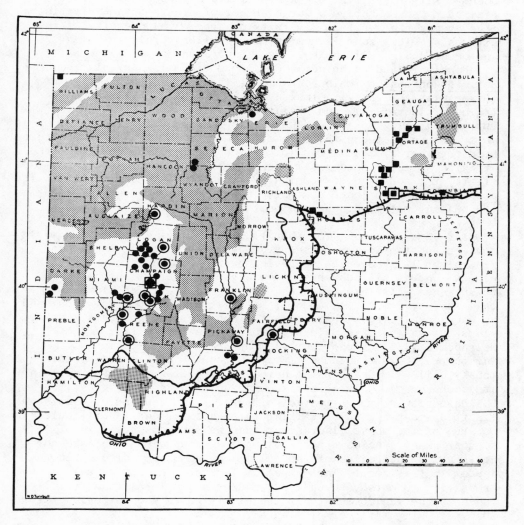

Fig. 1. Map of Ohio showing locations of 52 fens. Dots, extant
Prairie Fens; dots with circles, destroyed Prairie Fens; squares,
extant Bog Fens; square with open square, destroyed Bog Fen; large
dot with open square, Cedar Bog; solid line, maximum extent of
Wisconsinan glaciation; hashured line, maximum extent of continental Pleistocene glaciation; shaded area, maximum relief less
than 100 feet.

peat is developed which subsequently covers the marl. As this
sedge peat accumulates, it rapidly increases in thickness, often
becomes slightly elevated, and the terrain then becomes better
drained. Consequently, a large number of plants invade creating a
zone having the greatest diversity of species and density of individuals. This zone is dominated by taller herbaceous plants
and occasional small shrubs. Species common are Sanguisorba
canadensis (Canadian burnet), Cirsium muticum (swamp thistle),
Solidago ohioensis (Ohio goldenrod), Thelypteris palustris
(Dryopteris thelypteris) (marsh fern), Potentilla fruticosa
(shrubby cinquefoil), and Rhamnus alnifolia, (alder-leaved
buckthorn). The sedge-meadow zone is located adjacent to the open
marl zone and between the streamlets. Here the plates of peat are
held together by a multitude of fibrous roots suspended above the
artesian spring. Because of the loose arrangement of this substrate, the sedge-meadow community often quakes when walking on it
providing an experience similarly encountered when walking upon
the quaking mat in a sphagnum peat bog.

    At some distance from the artesian springs and the open marl
zone, a substantial increase develops in the amount of peat and
mineral soils. Here the sedge-meadow zone is replaced by a shrub-
meadow zone. This zone has a dense groundcover dominated by
tussocks of waist-high grasses and sedges, scattered clumps of
tall shrubs, especially alders, willows, and shrubby dogwoods, and
an occasional tree. Characteristic woody species of this zone are
Alnus rugosa (speckled alder), Aronia melanocarpa (Pyrus
melanocarpa) (black chokeberry), Physocarpus opulifolius
(ninebark), Cornus racemosa (gray dogwood), and Salix discolor
(pussy willow). Examples of herbaceous species are Carex stricta
(tussock sedge), Calamagrostis canadensis (blue-joint grass),
Solidago patula (rough-leaved goldenrod), and Eupatorium maculatum
(spotted joe-pye-weed). Walking through this zone is difficult
because of the dense plant cover and the numerous tussocks of
sedges. Many species typical of marshes and swamps that are not
necessarily indicative of the fen meadow community also begin to
appear in this zone. These species foretell the ecological
transition to marsh, shrub-carr, and/or swamp forest that totally
replaces the fen community if the highly calcareous artesian water
supply is diminished or eliminated.

                     HISTORY OF FLORISTIC STUDIES

    Because of the rare and unusual plants that occur in fens,
the pioneer Ohio botanists first visited them in the 1830's and
began recording the species encountered (Stuckey, 1966, 1974,
1978). John Samples, William S. Sullivant, Milo G. Williams, and

John W. VanCleve obtained plants from those fens near Dayton,
Springfield, and Urbana in west-central Ohio; Johnathan R. Paddock,
John L. Riddell, and William S. Sullivant from those near
Columbus in central Ohio; and Nicholas Riehl from those near
Canton in northeastern Ohio. It is somewhat difficult to discuss
the knowledge of fen habitats and the plants secured from them by
these botanists during these early times, because the term "fen"
was not in American botanical literature. Descriptions of fens
were included as a part of discussions on wet prairies, alkaline
bogs, peat bogs, marl meadows, or calcareous marshes. Using a
few key species, it is possible to identify and reconstruct
briefly the knowledge that the pioneer botanists had concerning
fen plants and their locations in Ohio. This reconstructed information has been useful in identifying the locations of fens
and has been incorporated on the map (Figure 1).

Beginning in the early 1890's, floristic surveys and ecological studies were conducted on selected alkaline wetlands in Ohio,
for example those at Cedar Bog (Cedar Swamp) in Champaign County
(Dachnowski, 1910; 1912, p. 39-43; Kellerman and Wilcox, 1895),
Castalia Prairie in Erie County (Moseley, 1899), and the Big
Spring Prairie in Wyandot County (Bonser, 1903). Extensive information on a number of calcareous peat deposits and alkaline
bogs in glaciated Ohio is provided by Alfred Dachnowski (1912) in
his classic book, Peat Deposits of Ohio. It was not until the
1940's, however, that the word "fen," a colloquial term referring
to plant communities of alkaline wetlands in Great Britain, was
applied to this type of vegetation and habitat in the United
States (Anderson, 1943). Since then interest in fens has been
increasing, and several fens in Ohio have been studied thoroughly
floristically by various students as a part of thesis work, but
little of their information has been published and correlated.
Since the 1930's published and manuscript information has become
available for the following fens in Ohio: Cedar Bog (Cedar
Swamp), Champaign County (Frederick, 1964, 1967, 1974a, b; King
and Frederick, 1974); Urbana Raised Bog, Champaign County (Gordon,
1933; McGill, 1971); Big Pond, Champaign County (McGill, 1973);
Sunbeam Prairie Bog, Darke County (Kapp and Gooding, 1964);
Castalia Prairie (Resthaven Wildlife Area), Erie County (Foos,
1971; Hurst, 1971; Sears, 1967); Pleasant Run Bog, Fairfield
County (Goslin, et al., 1970), Kick Fen and Lake Bend Fen,
Holmes County, (Wilson, 1974); Mud Lake Bog, Williams County
(Brodberg, 1976); Big Spring Prairie, Wyandot County (Stuckey,
1979). General descriptions of fens in Ohio have been written
by Gordon (1969) and Denny (1979). Other more general discussions
of fens, outside of Ohio, have been provided by Curtis (1959) for
Wisconsin, Moran (1981) for Illinois, Van der Valk (1975, 1976),
for Iowa, and Pringle (1980) for the Great Lakes region.

FLORISTIC SIMILARITIES AND DIFFERENCES

Our floristic studies have been conducted during the past 15 years in 30 or more fens in Ohio. For purposes of analysis, 20 fens (nine from west-central Ohio, seven from northeastern Ohio, two from northwestern Ohio, and one from south-central Ohio) and 74 species of vascular plants considered distinctive of fens were selected (Figure 2; Table 1). Of the 74 species, 30 of them, or approximately 40 percent, occur in most of the fens, whereas 24 species are primarily in the west-central fens and 20 species are more specific of the northeastern fens (Table 1). Those species in the west-central fens more typically occur in wet prairies of western Ohio, whereas those species in the northeastern fens are more typically associated with acid bogs in northern Ohio. Based on these floristic differences, the west-central fens are referred to as Prairie Fens and the northeastern fens are here designated as Bog Fens. The locations of the Prairie Fens (shown by dots and circles) and the Bog Fens (represented by open and closed squares) are on the accompanying map (Figure 1). Cedar Bog (Cedar Swamp) in Champaign County (located by a dot surrounded by an open square) is the largest, the most floristically diverse, and the most thoroughly studied of all the fens in Ohio. It is both a Prairie Fen and a Bog Fen, the only one in the state that contains a nearly full complement of both wet prairie and acid bog species. Of the 74 species analyzed, 63 occur there, including 21 of 24 wet prairie species, 12 of 20 acid bog species, and all 30 of the species characteristic of both types of fens. In northwestern Ohio, the fen at Castalia (Resthaven Wildlife Area) in Erie County is a Prairie Fen, and Mud Lake Bog in Williams County is a Bog Fen. In south-central Ohio, Blackwater Fen in Ross County is a Prairie Fen.

GEOGRAPHICAL AFFINITIES AND ORIGIN OF THE FLORA

Because the fens of Ohio are restricted to that portion of the state covered by the continental ice sheet during the most recent glacial period 15 to 20 thousand years ago, the geographical affinities and origin of this distinctive floristic element are of particular interest from an historical phytogeographic viewpoint. Botanists have long recognized that the flora of the northern hemisphere prior to glaciation consisted of many widely distributed northern or boreal circumpolar species that had developed and spread since the late Tertiary when climatic conditions were much warmer. With the onset of a cooler environment and extensive glaciation, these more or less continuous distribution patterns of the species of this Arcto-Tertiary flora were interrupted and became extensively modified by one or more periods of continental glaciation. This modification occurred to the

degree that only those species whose ranges either already extended far to the south in North America or whose plants were able
to migrate southward as environmental conditions changed with the
onset of glaciation were able to survive along the glacial ice
margin or in somewhat isolated areas away from the glacier. In
these isolated areas or survivia, called also refugia by some
authors, the surviving species or segrates of the species were
able to live at the southern extremes of their ranges. In the
southern part of North America, evidence exists for at least four
major geographical survivia--the Atlantic Coastal Plain, the
southern portion of the Appalachian Mountains, the Ozark Mountains,
and the central Rocky Mountains in western United States.

An examination of the present-day distribution patterns of
those northern Arcto-Tertiary species that are distinctive to Ohio
fens suggests their geographical affinity and origin. Based on
the observed geographical patterns, along with additional sup-
porting evidence from the morphology of the plants, proximity of
related species, and viewpoints of those individuals who have
studied the taxonomic and geographical relationships of these
species, an hypothesis can be developed for the origin of the fen
floristic element in Ohio. During continental glaciation none of
the Arcto-Tertiary species would have survived to the north of
Ohio; rather all of these species would have migrated within or
into the state from their locations near the ice margin or from
their survivia to the east or west. That is to say, a present-day
northern species whose distribution is nearly confined to the
glaciated territory, but with range extensions southward on the
Atlantic Coastal Plain and/or in the southern Appalachian Mountains
is considered of eastern affinity and probably invaded Ohio from
the east. Selected examples of species (Figures 3, 4, 5) mapped
with this distribution pattern are Sarracenia purpurea (pitcher
plant), Tofielda glutinosa (false asphodel), and Cladium
mariscoides (twig-rush). A northern species whose distribution
is nearly confined to the glaciated area, but with range extensions southward in the Ozark Mountains or central Rocky
Mountains is considered of western affinity and probably invaded
Ohio from the west. Selected examples of species or species complexes (Figures 6, 7, 8, 9) mapped with this distribution pattern
are Potentilla fruticosa (shrubby cinquefoil), Lobelia kalmii
(Kalm's lobelia), Valeriana ciliata (valerian), and Gentiana
procera (fringed gentian). Some species complexes (Figure 10)
have segregates living in more than one survivum, as is shown by
Zygadenus elegans in the Rocky Mountains and Z. glaucus (white
camass) in the Ozark Mountains and the Appalachian Mountains.
Species absent from the Rocky Mountains, but surviving in the
Ozark Mountains and the Appalachian Mountains are considered of
eastern affinity, and are represented by selected examples

Fig. 2. Map of Ohio showing locations of 20 fens whose flora is analyzed in this study.

Fen	County	Township
1. Castalia Prairie (Resthaven Wildlife Area)	Erie	Margaretta
2. Mickey Fen	Logan	Union
3. Urbana Raised Bog	Champaign	Urbana
4. Liberty Fen	Logan	Liberty
5. DeGraff Fen	Logan	Liberty
6. Buck Creek Fen (Buffenbarger Fen, Baldwin Fen)	Clark	Moorefield
7. Silver Lake Fen	Miami	Bethel
8. Prairie Road Fen	Clark	Moorefield
9. Zimmerman Prairie Fen	Greene	Beaver Creek
10. Blackwater Fen (Goodman Bog)	Ross	Green
11. Cedar Bog (Cedar Swamp)	Champaign	Urbana
12. Nimisilia Fen	Summit	Green
13. Kick Fen	Holmes	Washington
14. Frame Lake Bog	Portage	Streetsboro
15. Mud Lake Bog	Williams	Northwest
16. Streetsboro Bog	Portage	Streetsboro
17. Mantua Swamp	Portage	Mantua
18. Jackson Bog	Stark	Jackson
19. Wingfoot Lake	Portage	Suffield
20. Standard Slag Bog (Myersville Bog)	Summit	Green

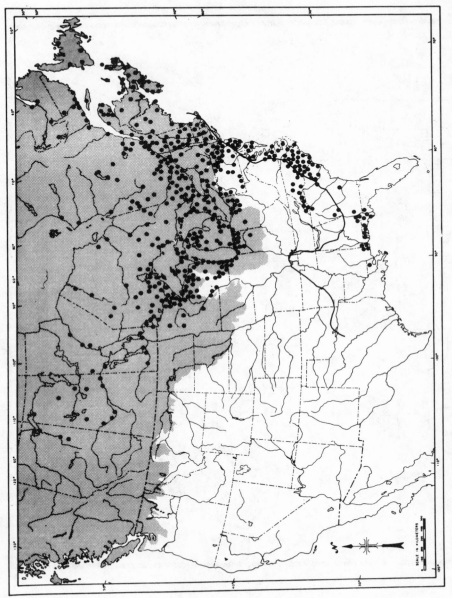

Fig. 3. Map of Sarracenia purpurea (pitcher plant) in North America, derived primarily from maps in Cody and Talbot (1973), Cruise and Catling (1971), McDaniel (1971), and Rousseau (1974).

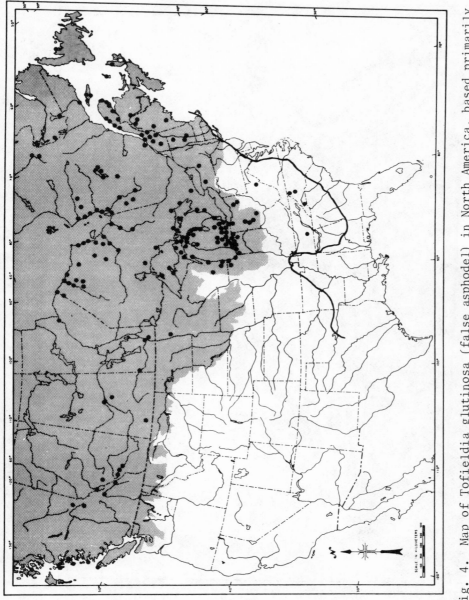

Fig. 4. Map of Tofieldia glutinosa (false asphodel) in North America, based primarily from maps in Hitchcock (1944), Lepage (1966), and Rousseau (1974).

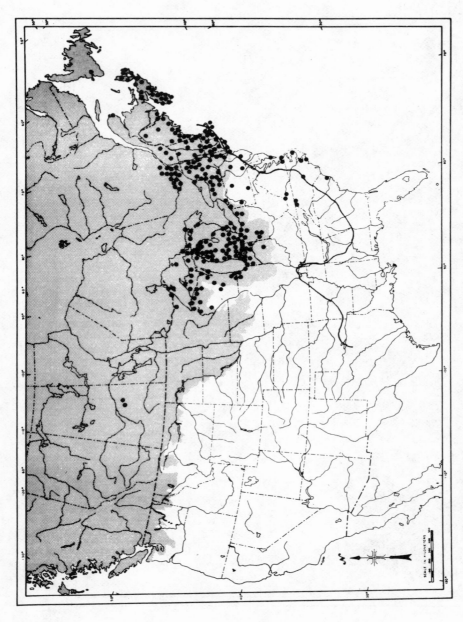

Fig. 5. Map of Cladium mariscoides (twig-rush) in North America derived primarily from maps in Maher et al. (1979), Raymond (1971), and Rousseau (1974) for the Canadian distribution.

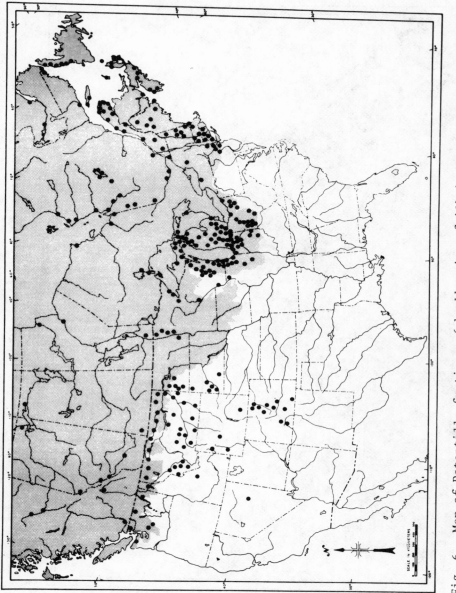

Fig. 6. Map of Potentilla fruticosa (shrubby cinquefoil) in North America, derived primarily from maps in Raup (1947) and Rousseau (1974) for the Canadian distribution.

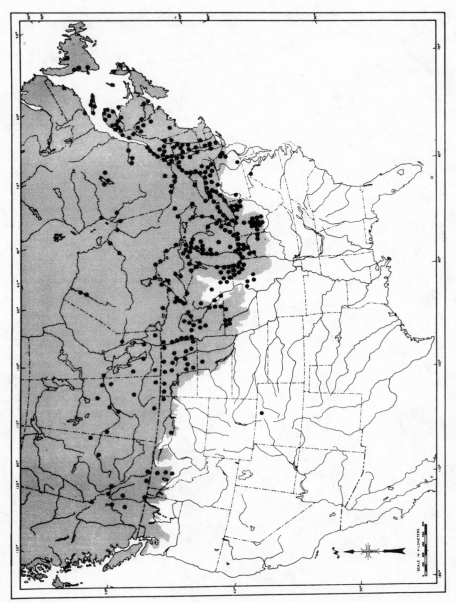

Fig. 7. Map of *Lobelia kalmii* (Kalm's lobelia) in North America, derived primarily from maps in Cody and Talbot (1978) and McVaugh (1936).

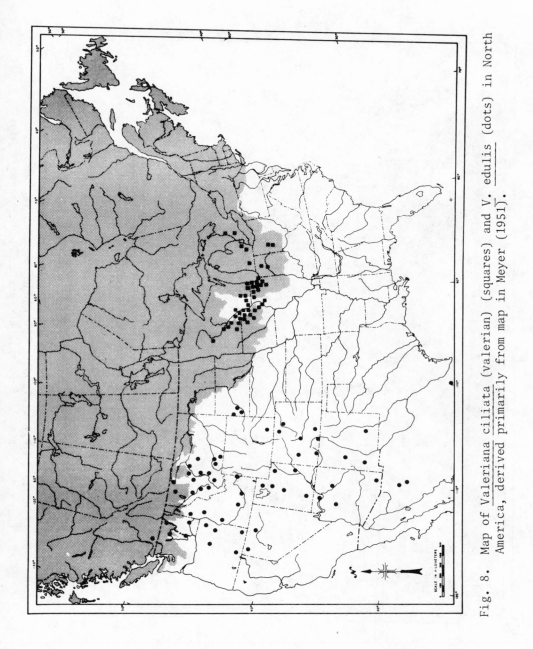

Fig. 8. Map of Valeriana ciliata (valerian) (squares) and V. edulis (dots) in North America, derived primarily from map in Meyer (1951).

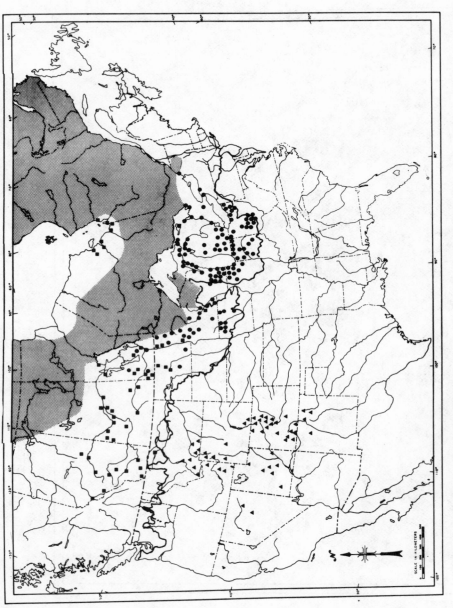

Fig. 9. Map of Gentiana procera (fringed gentian) (dots) and allied species, G. thermalis (triangles) and G. macounii (squares) in North America, derived from maps in Gillett (1957) and Iltis (1965). The shaded portion represents the area covered by the Laurentian Shield of sterile and acidic soils, taken from Iltis (1965).

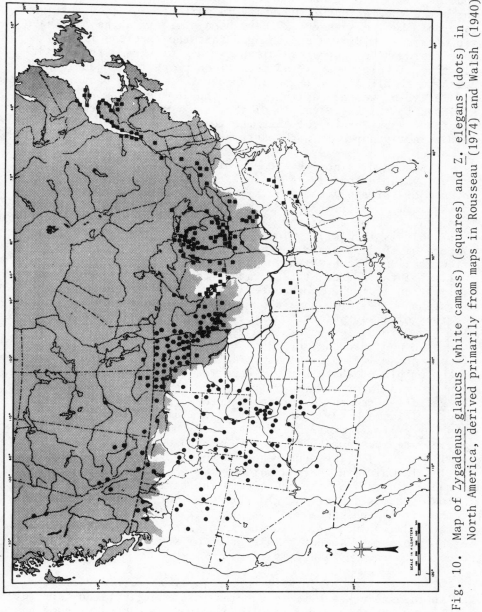

Fig. 10. Map of <u>Zygadenus glaucus</u> (white camass) (squares) and <u>Z. elegans</u> (dots) in North America, derived primarily from maps in Rousseau (1974) and Walsh (1940).

(Figures 11, 12, 13), such as Cirsium muticum (swamp thistle),
Cypripedium reginae (showy ladyslipper orchid), and Pedicularis
lanceolata (lousewort). The apparent absence of Arcto-Tertiary
species in areas between the survivia and the glaciated territory
where the species occur today may in part be explained that any
favorable habitats which may have been present in the past have
since been so modified and changed that the species, if they once
were there, have long since disappeared. Of the 74 species distinctive of Ohio fens, 49 of them or 66 percent are considered
northern (Table 1). Of these 49 species, 30 are of eastern affinity
and 19 are of western affinity. Of the 20 species mostly restricted to the Bog Fens, 19 of them are northern species. These
several statistics reveal that the fen floristic element in Ohio
is distinctly northern with eastern affinities.

     A second distinctive element of the fens in Ohio is repre-
sented by those species of southeastern United States, which are
not members of the Arcto-Tertiary flora. These species are
particularly well represented in the Prairie Fens of west-central
Ohio, where occur 20 of the 21 southeastern species selected for
this analysis (Table 1). These species are generally significant
members of the wet prairies in Ohio, and many of them have distributional patterns which suggest an origin from the southeastern
portion of the United States, rather than the more popular idea of
a migration from the southwestern part of the United States, a
migration believed to have occurred during the Xerothermic Period.
The idea of a southern or southeastern origin for the wet prairie
flora in Ohio or in the Prairie Peninsula as a whole was early
suggested and discussed by Adams (1920a, b; 1905) and later reviewed and amplified by Gleason (1923 p. 77). A further analysis
of the origin of the prairie floristic element in Ohio is deferred
as part of a later study. Of relevance here is that species such
as Agalinis purpurea (Gerardia purpurea) (gerardia), Liatris
spicata (blazing star), Scleria verticillata (nut-rush), and
Silphium trifoliatum (whorled rosinweed) belong to genera whose
species are most numerous in southeastern United States (Figures
14, 15, 16, 17). The occurrence of southeastern or wet prairie
species in Ohio fens have apparently come about from an extension
of their ranges from the south onto the newly exposed landscape
following recession of the glacier. However, southern species
of the Mississippi Embayment, typical of mudflats along river
shores and floodplains, evidently represent a distinctly different floristic element. This group of southern species is
absent from the fens in Ohio.

     The migration of formerly northern Arcto-Tertiary species
and southeastern wet prairie species into the glaciated territory
of Ohio probably occurred simultaneously with the melting of the
ice or shortly thereafter as the continental ice front receded.

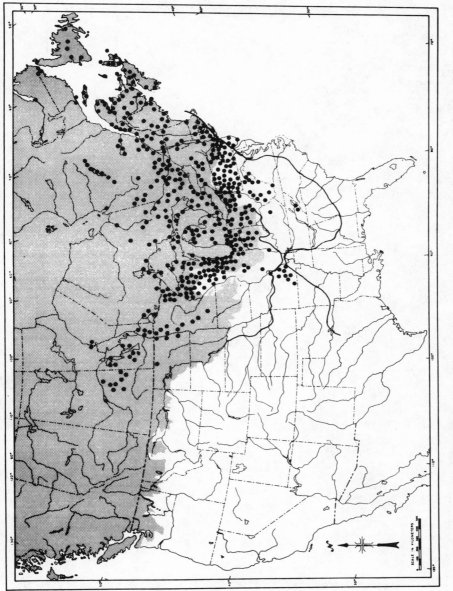

Fig. 11. Map of Cirsium muticum (swamp thistle) in North America, derived primarily from maps in Frankton and Moore (1963) and Rousseau (1974) for the Canadian distribution.

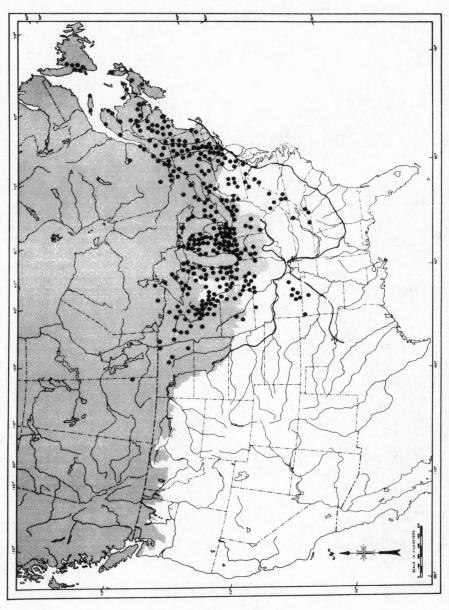

Fig. 12. Map of Cypripedium reginae (showy ladyslipper orchid) in North America, derived from maps in Case (1964), Fernald (1933), and Rousseau (1974).

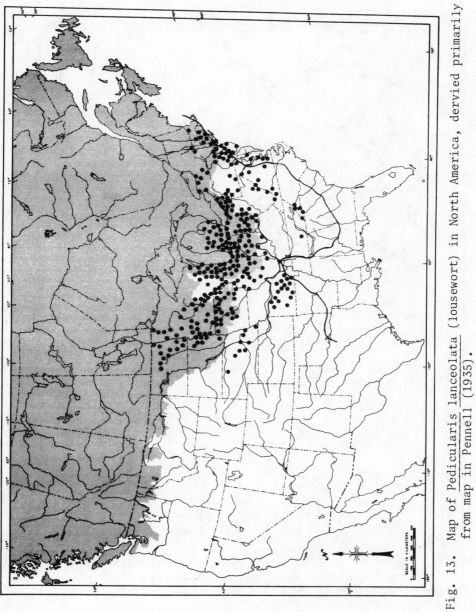

Fig. 13. Map of Pedicularis lanceolata (lousewort) in North America, dervied primarily from map in Pennell (1935).

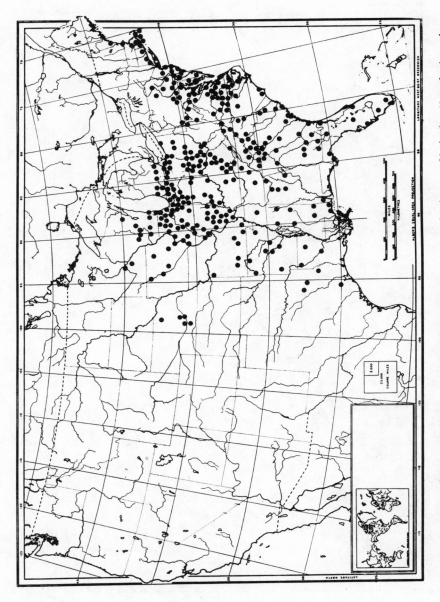

Fig. 14. Map of Agalinis purpurea (Gerardia purpurea) (gerardia) in North America, derived from map in Pennell (1935).

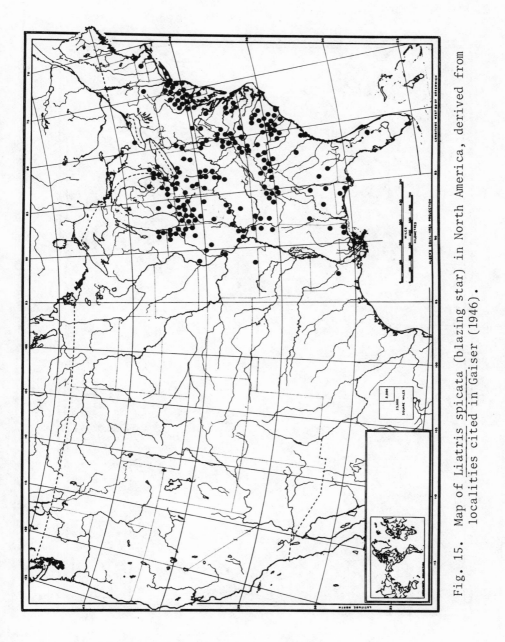

Fig. 15. Map of Liatris spicata (blazing star) in North America, derived from localities cited in Gaiser (1946).

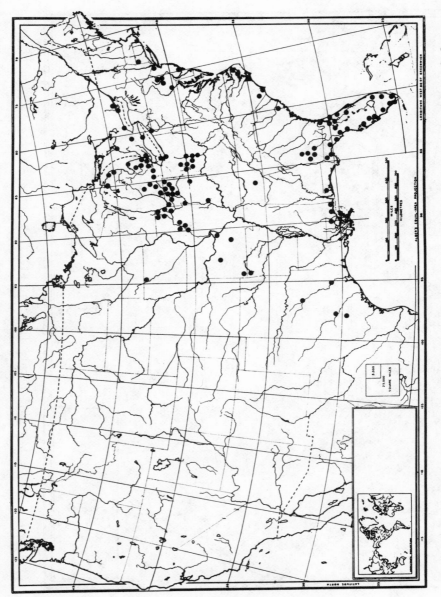

Fig. 16. Map of Scleria verticillata (nut-rush) in North America, derived from localities cited in Core (1936) and map in Fairey (1967).

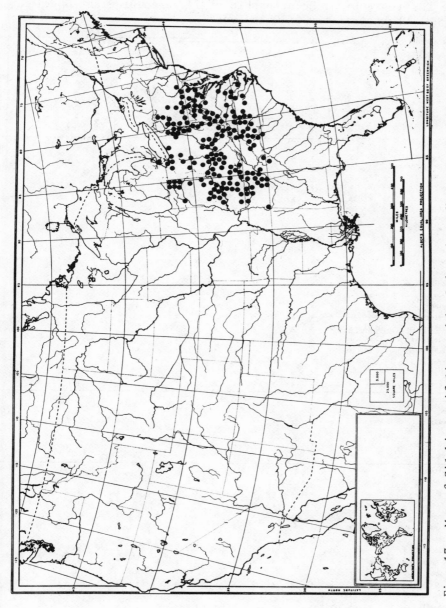

Fig. 17. Map of Silphium trifoliatum (whorled rosinweed) in North America, derived from map in Weber (1968).

Table 1. Geographical Relationship of 74 Vascular Plant Species Characteristic of Ohio Fens

Geographical Affinity and Number of Species	Prairie Fens →										← →	Bog Fens →								
	1	2	3	4	5	6	7	8	9	10	11	12	13	14	15	16	17	18	19	20
	Castalia Prairie	Mickey Fen	Urbana Raised Bog	Liberty Fen	DeGraff Fen	Buck Creek Fen	Silver Lake Fen	Prairie Road Fen	Zimmerman Fen	Blackwater Fen	Cedar Bog	Nimisilia Fen	Kick Fen	Frame Lake Bog	Mud Lake Bog	Streetsboro Bog	Mantua Swamp	Jackson Bog	Wingfoot Lake	Standard Slag Bog
Southeastern (20 species)																				
Agalinis purpurea (Gerardia purpurea)	x	x	x	x	x	x	x	x	x	x	x			x						
Allium cernuum	x	x	x	x	x	x		x			x									
Andropogon gerardii	x	x	x	x	x	x		x	x	x	x	x		x						x
Andropogon scoparius	x	x	x	x	x	x		x	x		x	x								
Cacalia plantaginea (C. tuberosa)			x			x	x	x	x	x										
Cyperus flavescens							x	x	x	x										
Filapendula rubra		x	x	x	x	x	x	x	x	x	x	x	x							
Helenium autumnale	x	x	x	x			x	x			x	x	x							
Juncus brachycephalus		x	x	x		x	x	x	x	x	x	x						x	x	
Liatris sqicata	x	x	x	x				x			x	x								
Lysimachia quadriflora	x	x	x	x	x	x	x	x	x	x	x	x	x							
Lythrum alatum	x		x	x		x	x	x	x	x										
Oxypolis rigidor	x	x	x	x	x	x					x	x	x	x	x	x				
Phystosegia virginiana		x	x	x	x	x	x	x			x									
Rhamnus lanceolata	x	x	x	x	x	x	x		x		x									
Scleria verticillata	x		x			x	x	x	x	x	x									
Silphium terebinthinaceum	x	x				x		x	x		x		x							
Silphium trifoliatum		x	x	x				x	x		x	x	x							
Solidago riddellii	x	x	x	x	x	x		x	x		x		x							
Sorghastrum nutans	x	x	x	x	x	x	x	x	x	x	x	x	x	x						
Northern (western) (4 species)																				
Eleocharis elliptica	x		x	x	x	x	x	x	x					x						
Juncus nodosus	x		x	x	x	x	x	x	x		x									
Triglochin maritimum			x	x				x			x									
Valeriana ciliata		x	x	x							x									
Northern (eastern) (17 species)																				
Aster puniceus	x	x	x	x		x	x	x	x	x	x	x	x	x	x	x	x	x	x	x
Carex flava		x		x		x		x			x							x		x
Cirsium muticum	x	x	x	x	x	x	x	x	x	x	x	x	x	x	x	x	x	x	x	x
Cladium mariscoides	x	x	x	x		x	x	x			x				x			x		x
Parnassia glauca	x	x	x	x	x		x	x	x	x	x	x		x		x	x	x		x
Pedicularis lanceolata	x	x	x			x	x	x	x	x	x	x	x	x	x	x	x			x
Rhynchospora capillacea	x	x		x		x	x	x	x	x	x	x		x		x		x		x
Physocarpus opulifolius	x	x	x				x				x		x							x
Pycanthemum virginianum	x	x	x	x	x	x	x	x	x	x	x	x	x					x	x	x
Sanguisorba canadensis		x	x			x	x	x			x	x	x	x				x	x	x
Solidago ohioensis	x	x	x	x		x	x	x	x	x	x	x		x				x	x	x

	1	2	3	4	5	6	7	8	9	10	11	12	13	14	15	16	17	18	19	20
Solidago uliginosa	x	x	x	x	x	x	x	x		x	x	x	x	x		x	x	x	x	x
Thelypteris palustris (Dryopteris thelypteris)	x	x	x	x		x	x	x		x	x	x	x	x	x	x	x	x	x	x
Tofieldia glutinosa		x		x	x			x			x	x		x				x		x
Viburnum lentago	x	x	x			x	x				x	x	x	x		x	x	x		x
Viola cucullata	x	x	x			x					x	x		x			x	x		x
Zygadenus glaucus		x	x	x				x			x			x			x	x		
Northern (western) (9 species)																				
Carex buxbaumii	x		x	x		x					x	x					x			
Eleocharis rostellata	x	x	x	x	x	x	x	x			x	x	x	x			x	x		x
Gentiana procera	x	x	x	x				x		x	x	x						x	x	x
Lobelia kalmii	x	x	x	x	x	x	x	x	x	x	x	x		x	x		x	x	x	x
Lycopus uniflorus	x	x	x	x	x	x	x	x	x	x	x			x	x		x	x	x	x
Muhlenbergia glomerata		x	x	x						x	x	x	x	x	x	x	x	x	x	x
Potentilla fruticosa	x	x	x	x	x	x	x	x	x	x	x	x	x	x	x	x	x	x	x	x
Scirpus acutus	x	x	x	x	x	x	x	x	x	x	x	x	x	x	x	x	x	x	x	x
Triglochin palustre	x			x			x	x	x		x							x		x
Other (4 species)																				
Eupatorium maculatum	x	x	x	x	x	x	x	x	x	x	x	x	x	x	x	x	x	x	x	x
Rudbeckia hírta	x	x	x	x	x	x	x	x	x	x	x	x	x	x	x	x				x
Scirpus americanus	x	x	x	x		x	x	x	x		x	x						x		
Thalictrum dasycarpum	x	x	x	x	x	x	x	x	x	x	x	x	x	x	x	x	x	v		x
Southeastern (1 species)																				
Calapogon tuberosus (C. pulchellus)	x										x	x	x				x			
Northern (eastern) (13 species)																				
Alnus rugosa											x	x	x	x	x	x	x	x	x	x
Aronia melanocarpa (Pyrus melanocarpa)											x			x	x	x	x	x		x
Betula pumila											x			x						
Cypridpedium reginae											x			x		x				
Drosera rotundifolia								x			x	x	x	x	x	x	x	x		x
Ilex verticillata											x			x						
Larix laricinia											x		x	x	x				x	x
Liparis loeseli				x							x	x		x						x
Myrica pensylvanica														x		x				
Pogonia ophioglossoides											x									
Rhus vernix							x				x	x	x	x	x	x	x	x	x	x
Sarracenia purpurea											x							x		x
Vaccinium macrocarpon														x	x		x			
Northern (western) (6 species)																				
Cypripedium candidum	x																x			
Eleocharis pauciflora								x			x			x						
Eriophorum virdi-carinatum											x			x		x	x		x	x
Rhamnus alnifolia											x			x		x	x	x		x
Salix candida	x											x		x	x	x		x		
Salix serrisima														x	x	x	x	x	x	
Totals	43	45	47	48	27	41	42	51	33	29	64	44	28	37	27	26	31	35	22	37

Northern Species Total 49 of 74 = 66%

While the glacier was melting, extensive portions of the landscape became highly disturbed with numerous newly created open, cool, wet habitats bare of vegetation. The cool meltwater would have been highly calcareous, particularly in certain areas in west-central Ohio, and to a lesser extent in northeastern Ohio. These extensive open, wet, naturally disturbed, calcareous habitats were ideal for the invasion of pioneer herbaceous and shrub species. The distinctive species of today's fens, in addition to many other pioneer species, undoubtedly invaded these new naturally disturbed sites in waves of migration that would have followed the line of retreat of the continential glacier. The distinctive fen species in those days would probably have been much more plentiful, where they would have grown along the open, calcareous, cool meltwater streams in Ohio. However, as the meltwater supply subsided, the soil conditions modified, and the landscape became warmer and drier, other species of later suc-cessional stages invaded and became the dominant vegetation. The pioneer distinctive species of the fens were able to persist only at those sites where a supply of cool calcareous ground water continued to flow from artesian springs creating marl flats that provided a disturbed habitat necessary for their maintenance. Only where this type of habitat has been maintained in Ohio are the fens located today. In these sites the distinctive pioneer species of the fens are still maintained as relict fragile plant communities persisting from the time of glaciation 15 to 20 thousand years ago. Man's destructive activities of only 200 years have already destroyed nearly one-fourth of the 52 known fens in Ohio. The protection of those fens remaining must be secured if this floristic element, a remnant from the past ice age, is to continue to survive in Ohio.

REFERENCES

Adams, C. C. 1902a. Southeastern United States as a center of geographical distribution of flora and fauna. Biol. Bull. 3: 115-131.

-------. 1902b. Postglacial origin and migrations of the life of the northeastern United States. J. Geogr. 1: 303-310, 352-357.

-------. 1905. The postglacial dispersal of the North American biota. Biol. Bull. 9: 53-71.

Anderson, W. A. 1943. A fen in northwestern Iowa. Amer. Midl. Naturalist 29: 787-791.

Bonser, T. A. 1903. Ecological survey of Big Spring Prairie, Wyandot County, Ohio. Ohio Acad. Sci. Spec. Pap. No. 7. 99 pp.

Brodberg, R. K.   1976.   Vascular macrophytes of Mud Lake, Williams
    County, Ohio.   M.S. Thesis, Bowling Green State Univ.,
    Bowling Green, Ohio.   68 pp.

Curtis, J. T.   1959.   Fen, meadow and bog, pp. 361-384.   In   The
    vegetation of Wisconsin: An ordination of plant communities.
    Univ. Wisconsin Press, Madison.   xi + 657 pp.

Dachnowski, A.   1910.   A cedar bog in central Ohio.   Ohio Nat.
    11:   193-199.   (Reprinted, 1974.   C. C. King and C. M.
    Frederick, eds.   Cedar Bog Symposium Urbana College November
    3, 1973, Ohio Biol. Surv. Circ. No. 4.   pp. 55-59).

-------.   1912.   Peat deposits of Ohio: their origin, formation
    and uses.   Bull. Geol. Surv. Ohio 4th Ser. 16: 1-424.

Denny, G. L.   1979.   Relicts of the Past--Bogs, pp.141-150.   In
    M. B. Lafferty, ed. Ohio's Natural Heritage.   Ohio Acad.
    Sci., Columbus, Ohio.   324 pp.

Foos, K. A.   1971.   A floristic and phytogeographic analysis of
    the fen element at the Resthaven Wildlife Area (Castalia
    Prairie), Erie County, Ohio.   M.S. Thesis, The Ohio State
    Univ., Columbus, 81 pp.

Frederick, C. M.   1964.   Natural history study of the flora of
    Cedar Swamp.   I.   The bog meadow.   M.S. Thesis, The Ohio
    State Univ., Columbus.   102 pp.

-------.   1967.   A natural history survey of the vascular flora
    of Cedar Bog, Champaign County, Ohio.   Ph.D. Dissertation,
    The Ohio State Univ., Columbus.   240 pp.

-------.   1974a.   Disjunct plant species in Cedar Bog.   pp. 16-20.
    In C. C. King and C. M. Frederick, eds.   Cedar Bog Symposium
    Urbana College November 3, 1973.   Ohio Biol. Surv. Inform.
    Circ. No. 4.   71 pp.

-------.   1974b.   A natural history of the vascular flora of
    Cedar Bog, Champaign County, Ohio.   Ohio J. Sci. 74: 65-116.

Gleason, H. A.   1923.   The vegetational history of the Middle
    West.   Ann. Assoc. Amer. Geogr. 12: 39-85.   (Contrib.
    N.Y. Bot. Gard. No. 242).

Goldthwait, R. P., G. W. White, and J. L. Forsyth. 1961. Glacial
    map of Ohio. Department of the Interior, U. S. Geol. Surv.
    Prepared in cooperation with Div. Water & Div. Geol. Surv.,
    Ohio Dept. Nat. Resources, Columbus. Misc. Geol. Invest.
    Map 1-316.

Gordon, R. B. 1933. A unique raised bog at Urbana, Ohio. Ohio
    J. Sci. 33: 453-459.

-------. 1969. Freshwater marshes and fens, pp. 64-66. In
    R. B. Gordon. The natural vegetation of Ohio in pioneer
    days. Bull. Ohio Biol. Surv. n. ser. 3(2): i-xi, 1-113.

Goslin, R. M., W. E. Goslin, and C. R. Goslin. 1970. Flora of
    Pleasant Run Bog, section nine Berne Township, Fairfield
    County, Ohio. Typewritten. 6 pp.

Hurst, S. J. 1971. Geographical relationships of the prairie
    flora element and floristic changes from 1890-1970 at the
    Resthaven Wildlife Area (Castalia Prairie), Erie County,
    Ohio, with an appended list of vascular plants. M.S.
    Thesis, The Ohio State Univ., Columbus, 177 pp.

Kapp, R. O., and A. M. Gooding. 1964. A radiocarbon-dated pollen
    profile from Sunbeam Prairie Bog, Darke County, Ohio. Amer.
    J. Sci. 262: 259-266.

Kellerman, W. A., and E. M. Wilcox. 1895. First list of plants
    of Cedar Swamp, Champaign County, Ohio. 3rd Ann. Rep. Ohio
    Acad. Sci. 27-28.

King, C. C., and C. M. Frederick, eds. 1974. Cedar Bog
    Symposium Urbana College November 3, 1973. Ohio Biol.
    Surv. Inform. Circ. No. 4. 71 pp.

McGill, N. R. 1971. The Urbana Raised Bog - forty years later.
    Typewritten. 11 pp.

-------. 1973. A comparison of the vascular plant flora of two
    lakes in northern Champaign County, Ohio. M.S. Thesis,
    The Ohio State Univ., Columbus. 59 pp.

Moran, R. C. 1981. Fens in northeastern Illinois: Floristic
    composition and disturbance, pp. _____. In R. L. Stuckey,
    and K. J. Reese, eds. The Prairie Peninsula--in the
    "shadow" of Transeau: Proceedings of the Sixth North
    American Prairie Conference, The Ohio State Unv., Columbus,
    Ohio, 12-17 August 1978. Ohio Biol. Surv. Notes
    No. 15. _____ pp.

Moseley, E. L. 1899. Sandusky Flora. A catalogue of the
    flowering plants and ferns growing without cultivation in
    Erie County, Ohio and the peninsula and islands of Ottawa
    County. Ohio State Acad. Sci. Spec. Pap. No. 1. 167 pp.

Pringle, J. S. 1980. An introduction to wetland classification
    in the Great Lakes region. Royal Bot. Gardens Tech. Bull.
    No. 10. 11 pp.

Sears, P. B. 1967. The Castalia Prairie. Ohio J. Sci. 67:
    78-88.

Stuckey, R. L. 1966. The botanical pursuits of John Samples,
    pioneer Ohio plant collector (1836-1840). Ohio J. Sci.
    66: 1-41.

-------. 1974. Early botanical exploration of Cedar Bog.
    pp. 21-24. In C. C. King and C. M. Frederick, eds. Cedar
    Bog Symposium Urbana College November 3, 1973. Ohio Biol.
    Surv. Inform. Circ. No. 4. 71 pp.

-------. 1978. Medical botany in the Ohio Valley (1800-1850).
    Trans. and Stud. Coll. Physicians Philadelphia 45: 262-279.

-------. 1979. A checklist of vascular plants of Big Spring
    Prairie taken from Bonser (1903) with revised scientific
    names. Typewritten. 8 p.

Wilson, H. D. 1974. Vascular plants of Holmes County, Ohio.
    Ohio J. Sci. 74: 277-281. Catalog of Vascular Plants.
    Typewritten, Mimeographed. 55 pp.

Van der Valk, A. G. 1975. Floristic composition and structure
    of fen communities in northwest Iowa. Proc. Iowa Acad.
    Sci. 82: 113-118.

-------. 1976. Zonation, competitive displacement and standing
    crop of northwest Iowa fen communities. Proc. Iowa Acad.
    Sci. 83: 51-54.

SOURCES OF INFORMATION FOR THE DISTRIBUTION MAPS

The information used in developing the plant distribution
maps has been obtained from many sources too numerous to cite
for each individual species. The principle reference(s) for
each map is provided in the figure caption (references cited
below). Other sources are the maps cited in Index to Plant
Distribution Maps in North American Periodicals through 1972 by
W. L. Phillips and R. L. Stuckey (1976, G. K. Hall & Co.,

Boston, xxxvii + 686 pp.), state and local floras which have maps
and/or cited localities, atlases of plant distribution maps, state
rare and endangered species reports, and herbarium records--
principally those seen at MICH and OS.

Case, F. W., Jr.   1964.   Orchids of the western Great Lakes
    region.   Cranbrook Institute Sci. Bull. 48.   147 pp.

Cody, W. J., and S. S. Talbot.   1973.   The pitcher plant
    Sarracenia purpurea L. in the northwestern part of its
    range.   Canad. Field-Naturalist 87: 318-320.

--------.   1978.   Vascular plant range extensions to the Heart
    Lake area, District of Mackenzie, Northwest Territories.
    Canad. Field-Naturalist 92: 137-143.

Core, E. L.   1936.   The American species of Scleria.   Brittonia 2:
    1-105 + pls. 1-3.

Cruise, J. E., and P. M. Catling.   1971.   The pitcher-plant in
    Ontario.   Ontario Naturalist 1971: 18-21.

Fairey, J. E., III.   1967.   The genus Scleria in the southeastern
    United States.   Castanea 32: 37-71.

Fernald, M. L.   1933.   Recent discoveries in the Newfoundland
    flora.   Part I.   Journal of 1926 and 1929.   Rhodora 35:
    1-16, 47-63, 80-95.

Frankton, C., and R. J. Moore.   1963.   Cytotaxonomy of Cirsium
    muticum, Cirsium discolor, and Cirsium altissimum.   Canad.
    J. Bot. 41: 73-84.

Gaiser, L. O.   1946.   The genus Liatris.   Rhodora 48: 165-183,
    216-263, 273-326, 331-382, 393-412.

Gillett, J. M.   1957.   A revision of the North American species
    of Gentianella Moench.   Ann. Missouri Bot. Gard. 44: 195-269.

Hitchcock, C. L.   1944.   The Tofieldia glutinosa complex of western
    North America.   Amer. Midl. Naturalist 31: 487-498.

Iltis, H. H.   1965.   The genus Gentianopsis (Gentianaceae):
    Transfers and phytogeographic comments.   Sida 2: 129-154.

Lepage, E.   1966.   Apercu floristique de secteur nord-est de
    l'Ontario.   Naturaliste Canad. 93: 207-246.

Maher, R. V., G. W. Argus, V. L. Harms, and J. H. Hudson. 1979.
    The rare vascular plants of Saskatchewan. Syllogeus No. 20.
    55 pp. + maps.

McDaniel, S. 1971. The genus Sarracenia (Sarraceniacea). Bull.
    Tall Timbers Research Station 9: 1-36.

McVaugh, R. 1936. Studies in the taxonomy and distribution of
    the eastern North American species of Lobelia. Rhodora 38:
    241-263, 276-298, 305-329, 346-362.

Meyer, F. G. 1951. Valeriana in North America and the West
    Indies (Valerianaceae). Ann. Missouri Bot. Gard. 38:
    377-503.

Pennell, F. W. 1935. The Scrophulariaceae of eastern temperate
    North America. Acad. Nat. Sci. Philadelphia Monogr. No. 1.
    650 pp.

Raup, H. M. 1947. The botany of southwestern Mackenzie.
    Sargentia 6: 1-275 + XXXVII Pls.

Raymond, J. 1971. Distribution Canadienne du Cladium
    mariscoides (Muhl.) Torr. Naturaliste Canad. 98: 735-737.

Rousseau, C. 1974. Geographie Floristique du Quebec-Labrador:
    Distribution des Principales Especes Vasculaires. Les
    Presses de l'Universite Laval, Quebec. 798 pp.

Walsh, O. S. 1940. A systematic study of the genus Zigadenus
    Michx. Ph.D. Dissertation, Univ. California, Berkeley.
    137 pp.

Weber, W. R. 1968. Biosystematic studies in the genus Silphium
    L. (Compositae). Investigations in the selected intra-
    specific taxa of Silphium asteriscus L. Ph.D. Dissertation,
    The Ohio State Univ., Columbus. 209 pp.

THE PALEOECOLOGICAL EVIDENCE FOR ENVIRONMENTAL CHANGES IN "NEOPALEOBOTANICAL" SEDIMENTS OF SOUTH FLORIDA

Peter R. Kremer and W. Spackman

The Pennsylvania State University

University Park, Pennsylvania

ABSTRACT

Eleven tree island communities (hammocks) in the Everglades of southern Florida were analyzed with regard to their vegetation, petrology of subsurface peats, and bedrock relationships. Fifteen phyteral-based peat types were reduced to three types based on water regime. The sequence of Holocene peats seen in the cores reflect a hydrosere successional sequence from basal hydric, to mesic-hydric, and ultimately, to mesic environments at the modern surface. Similar petrologically based analyses on solid (and semi-solid) natural organic deposits can be made for any age materials.

INTRODUCTION

In an effort to elucidate the nature of changes associated with the conversion of plant materials to peat (and then to coal), use has been made of peat petrology. In order to follow this transition in detail, the distinctive standing vegetation and community structure of some tree islands (hammocks) in the Everglades of south Florida (a potentially coal-forming wet-lands ecosystem) were characterized and the underlying peats studied as a means of reconstructing paleoecological events in the basin of deposition. Figure 1 identifies the location of the study site.

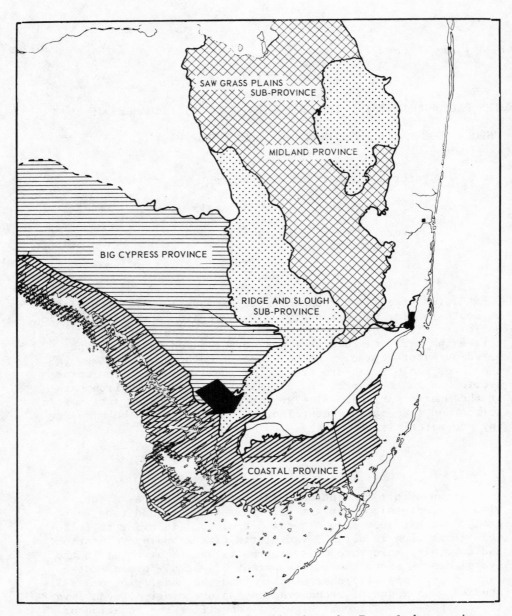

Fig. 1. Map of southern Florida, showing the Everglades region,
         its major peat provinces, and location of study site
         (arrow).

Fig. 2. A typical tree island (hammock) of the Everglades. They are generally oriented North-South.

These communities are commonly called hammocks (or tree islands): the one under detailed study is designated at BUP-4L-126 (from the aerial photo code number). This is one of many similar structures in this telmatic milieu and they all rest upon more or less substantial peat bodies which occupy localized bedrock lows in the carbonate substrate. Figure 2 shows the general morphology and physiography of one of these vegetational communities. Their head (northerly) ends are usually arboreal and the tree islands attenuate into brushy and, ultimately, herbaceous areas at their tail ends. These merge with the herbaceous marshes which surround these communities.

Each type of community produced a deposit of accumulated organic and inorganic vegetable materials which matured into peat. Petrographic examination of the peats revealed that each community, or sub-community, produced a respectively discriminate peat type. The types are so distinctive that transitions between them are readily evident. The Holocene peats were analyzed quantitatively and qualitatively, using their phyterals to characterize them. The sequences of peats were represented by approximately fifteen phyteral-based types; a phyteral being a petrologically recognizable fragment of a plant tissue or organ. However, when observed from the viewpoint of the modern water regime in which they developed, there were only three types indicated: 1) the hydric types (the wettest), 2) the mesic types (the driest), and 3) the hydric-mesic types (the water regime was intermediate to the two extremes). The hydric peat types are those derived from water lily and its phytic associates. The mesic peat types are those derived mostly from brushy and arboreal species within the swamp. The hydric-mesic peat types are those derived primarily from the marshy phytic elements comprising their respective communities.

The observed vegetational and peat sequences made possible paleoecological interpretations as far back as approximately 5000 year B.P. Data from others working to the north and to the south of the study site have established ^{14}C dates which range from approximately 4500-5000 years B.P. (see Smith, 1968). Therefore, the assumption is made that the basal peats in the deepest peat deposit of the hammocks studied are no older than 5000 years B.P.

METHODS

A transect line was established through the tree island BUP-4L-126, perpendicular to its long axis; along this line depth probes were made at 10 foot (3.05 meters) intervals. Species of vegetation along this transect were tabulated and

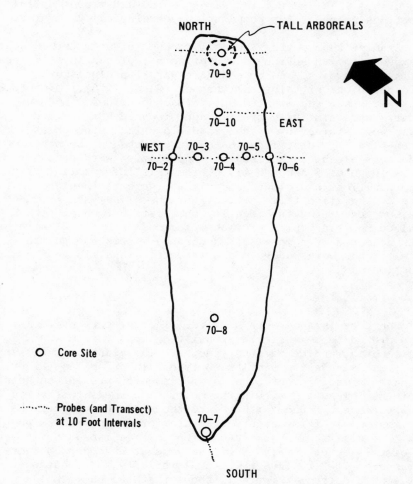

Fig. 3. Diagram of tree island BUP-4L-126 showing core sites
         and sampling transects.

the data analyzed, and three-inch (7.5 centimeter) diameter
cores were taken (Figure 3). The depth-probe data (using the
water level as datum line) provide a profile of the bedrock and
of the peat mass resting upon it (Figure 4). Data from vegetational analyses serve to characterize the community and
delineate such zonation in the standing vegetation as is present
at the surface. The peat cores were sampled and microscopical
thin sections prepared for comparative petrological analysis.

To examine them petrographically, cubes of undisturbed peat
were removed from three-inch diameter cores, placed in stainless
steel tissue capsules, and embedded in paraffin blocks by
standard botanical histological methods (Gabe, 1976). Microtomed
sections were made and mounted on microscope slides. Modern
plant parts were similarly treated and prepared but the modern
materials were stained by established biological procedures
(Johansen, 1940). Identification of structures in the peats
was made possible by comparison and correlation with modern
materials (Figures 5 through 11). From this information, past
plant communities, hydrology, and climate were definable in
absolute and relative terms. Data from these analyses provide
information concerning the identity and distribution of vegetational communities which previously occupied the local site.
This was achieved by comparison, under the microscope, of sub-
fossil materials with modern plant structures and tissues in
thin section.

Modern Vegetation

Eleven tree islands were sampled. The vegetational communities proved to be variable, consisting mostly of mixed
arboreal, shrub, and herbaceous species. The plant communities
of most tree islands are similar to tree island 73-KT-4 (Figure
12) which, typically zonate, has hydric species present on both
sides of the island and within the moat. Intermediate (hydric-
mesic) species are found farther into the stand of vegetation,
and the most mesic forms occupy the interior or interior high
points. The distribution of vegetation in tree island BUP-4L-
126 is of this type. Analysis of the vegetation along an east-
west transect (70-2 to 70-6, Figure 3) reflects the zonation to
a considerable degree. Figure 12 shows that the hammock is
bordered by aquatics such as Mariscus jamaicensis (Crantz) Britt.
(sawgrass), Bacopa caroliniana (Walt.) Robins. (aromatic
figwort), Leersia spp. (rice cut-grass), Pontederia lanceolata
Nutt. (southern pickerelweed), Pluchea spp. (fleabane), and
Annona glabra L. (pond apple). Inspection of the plant
community outside of the immediate line of the transect revealed

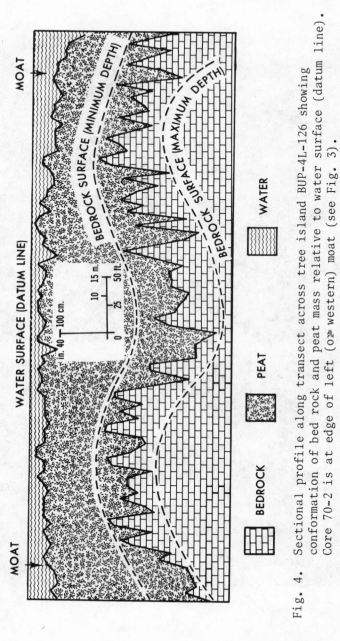

Fig. 4.   Sectional profile along transect across tree island BUP-4L-126 showing
conformation of bed rock and peat mass relative to water surface (datum line).
Core 70-2 is at edge of left (or western) moat (see Fig. 3).

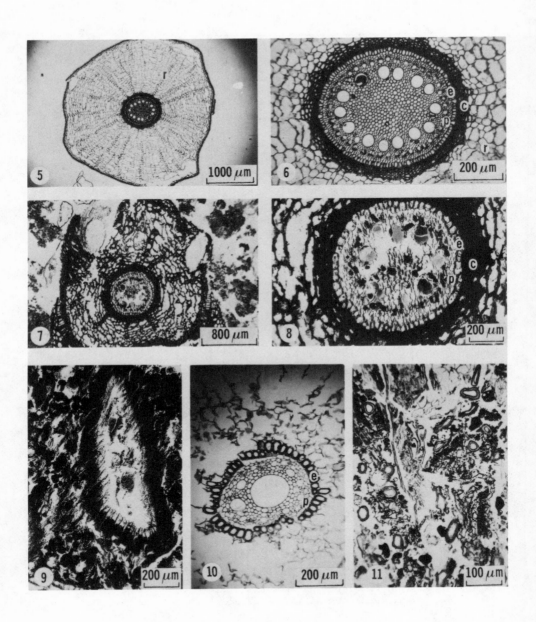

Fig. 5.   Modern *Mariscus jamaicensis* root, cross section; note
          radial lines of thick-walled cortical parenchyma (r).

Fig. 6.   Modern *Mariscus jamaicensis* root, cross section of
          stelar area; note endodermal (e) and pericyclic (p)
          layers surrounded by massive inner cortical tissue (c)
          with radiating lines of thick-walled cortical
          parenchyma (r).

Fig. 7.   Sub-fossil *Mariscus jamaicensis* root, cross section;
          from upper 25 cm of peat. Note radial lines of
          thick-walled cortical parenchyma and compare general
          tissue organization with modern specimen (Fig. 6).

Fig. 8.   Sub-fossil *Mariscus jamaicensis* root, cross section;
          from upper 25 cm. of peat. Compare endodermis,
          pericycle, and massive inner cortical tissue with
          counterparts in Fig. 7.

Fig. 9.   Sub-fossil *Mariscus jamaicensis* root, cross section;
          sampled below 25 cm. in peat. Note progressive
          degradation of all tissues; relative position of
          degraded tissues and cell masses permit identification
          of this phyteral.

Fig. 10. Modern *Sagittaria* spp. root, cross section of stelar
          area; note endodermal and pericyclic tissues.

Fig. 11. Sub-fossil *Sagittaria* spp. root; from below 25 cm. in
          depth. Endodermal cells are disaggregated and dis-
          persed; inner stelar tissues degraded beyond
          recognition.

the presence of Rhynchospora spp. (beak rush), Nymphaea odorata
Ait. (white water lily), Nymphoides aquatica (J. F. Gmel.)
Kuntze (floating hearts), Chara spp. (stoneworts), and
Utricularia spp. (bladderworts) (Long and Lakela, 1971).

This zonation is further reflected in the distributions seen
in Figure 14. The hydric-mesics are well distributed along the
entire transect with a better representation seen along the ends
of the transect. These species include Salix carolinana Michx.
(southern willow), Myrica cerifera L. (wax myrtle), Cephalanthus
occidentalis L. (buttonbush), Boehmeria cylindrica (L.) Willd.
(button hemp), an unknown "aroid", and numerous unidentified
seedlings. Outside the transect there were seen occasional
representatives of the hydric species.

The distribution of mesic species is seen in Figure 15.
These include Persea borbonia (L.) Spreng. (red bay), Thelypteris
palustris Schott (southern swamp fern), Blechnum serrulatum
Rich (swamp fern), Rapanea guianensis Aubl., unidentified grasses,
and a recurring unidentified sedge. Outside the transect were
such species as Magnolia virginiana L., Cyperus spp., and
Parthenocissus quinquefolia (L.) Planch. (Virginia creeper).

Peat Petrology

Petrological materials were prepared using a method modified from Radforth and Eydt (1958); point counts were made as
recommended by the International Committee for Coal Petrology
(1963). Cores were taken at sites indicated in Figure 3.

A portion of the assembled data is seen in Figures 16 and
17. Core 70-9 lies at the head end of the island and core 70-7
lies at the tail end; they lie approximately on the longitudinal
axis of the tree island. In longitudinal profile (Figure 16),
it can be seen that the peat body thins from its head toward
its tail. The vertical succession of peat types (as seen in
Core 70-9) indicates initial deposition of basal hydric peats,
a transition to hydric-mesic peats, and finally, mesic peats
topping out the peat column. Spatially, the surface vegetation
and sediments indicate a trend from a woodier, or mesic, element
at the head of the tree island to the development of less woody
and more herbaceous peats in the direction of the tail. The
oldest peats, those at the bottom of 70-9, are those of a hydric
nature; the youngest peats, those at the surface, are those
least hydric (mesic). From this observation a corollary follows:
developmentally (or successionally) the arboreal community at

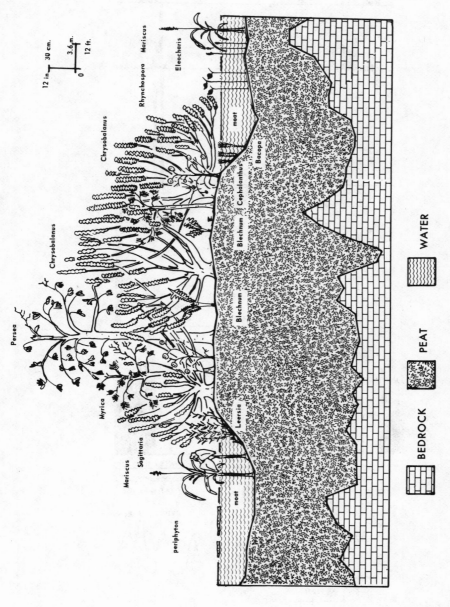

Fig. 12. Sectional profile and diagrammatic representation of vegetation along transect across tree island 73-KT-4 showing zonation of the tree island (hammock) community.

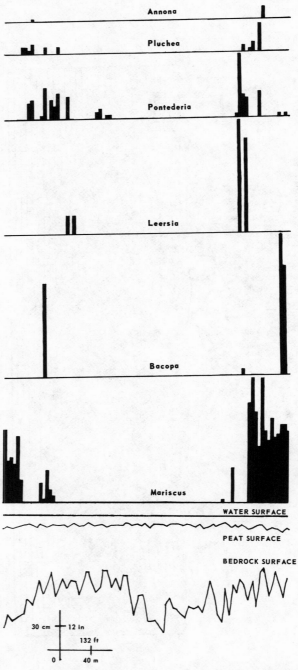

Fig. 13.   Distribution of hydric plant genera along the transect
across tree island (hammock) BUP-4L-126.

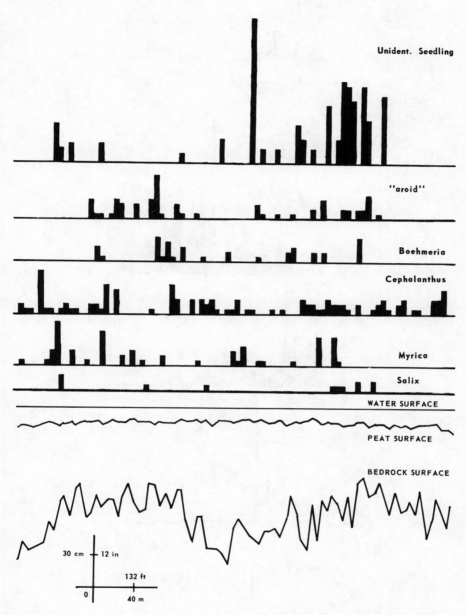

Fig. 14.   Distribution of hydric-mesic plant genera along the
           transect across tree island (hammock) BUP-4L-126.

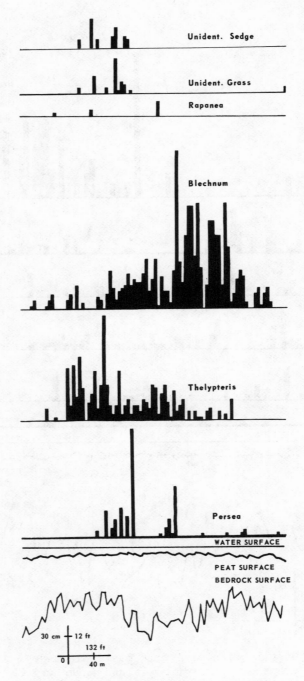

Fig. 15. Distribution of mesic plant genera along the transect
across tree island (hammock) BUP-4L-126.

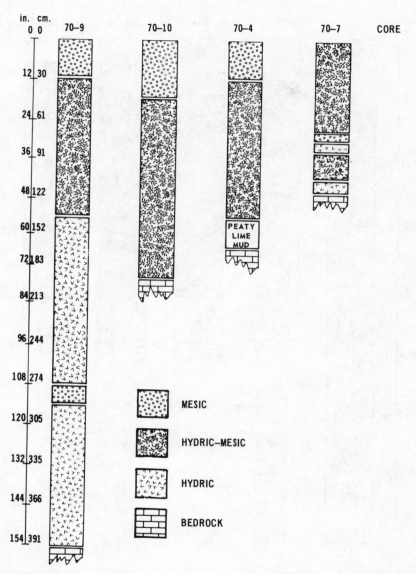

Fig. 16. Peat types in cores taken along longitudinal axis of
          tree island (hammock) BUP-4L-126.

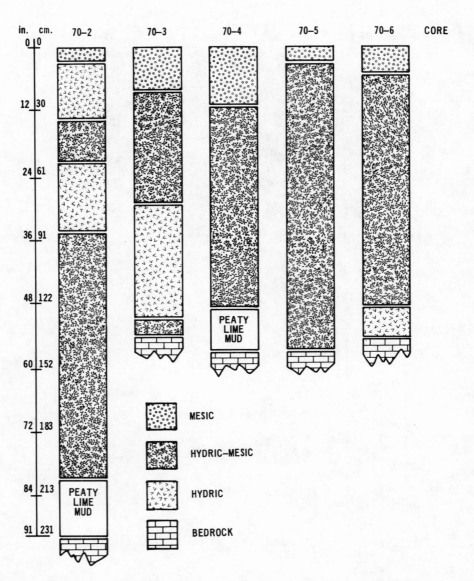

Fig. 17.   Peat types in cores taken along transect (cross-section) of tree island (hammock) BUP-4L-126.

the head end is an older community than the non-arboreal one at
the tail end of the tree island. In other words, within the
framework of community maturation, the community of the head end
of the tree island is more mature than the pioneer hydric community of the tail end.

A major point to be made is that Walther's principle of
sedimentary sequences is clearly demonstrated here. The
vertical sequence of deposits (and the distinctive environments
represented in their formation) is visible horizontally at the
surface.

In the cross-sectional profile (Figure 17), much the same
situation obtains. Core 70-4 occupies the middle position on
the transect and is on the approximate longitudinal axis; cores
70-2 and 70-6 occupy the margins of the tree island at the inner
edge of the moat, while 70-3 and 70-5 are spaced equidistant
from their neighboring cores (see Figure 3). Here, the general
sequence of events suggests hydric peats as basal, with the
least hydric peats forming the surface peats. (The significance
of the peat sequence in 70-2 must be judiciously examined because
it lies adjacent to a deepening bedrock low on the left margin
of the tree island [Figure 4]). In comparing the near-surface
peats of the five cores, the greater quantity of mesic peats in
the central area of the island indicates a higher degree of
maturation for the middle of the tree island, the least mature
areas are those at the borders. This strongly suggests a centri-
fugal mode of maturation of the vegetational community.

SUMMARY AND CONCLUSIONS

As seen from the distribution of peat types found in the
cores, the water levels were relatively high at the outset of
development on the newly emerged bedrock. The organisms comprising the peats were typical of those of a shallow pond or
small lake. This model fits the general bedrock profile from
which the samples were taken. As the basin filled in, the
community changed to one less and less hydric with the culmination
of the present day mesic, subclimax community.

The data, therefore, indicate a (vegetational) community
maturation sequence which proceeds, in two dimensions, from the
head end of the tree island toward the tail end of the tree
island and, simultaneously, is accompanied by lateral growth and
development from the center toward its outer edges. This results
in a tear-drop shape with attentuation toward the downstream end.
A third dimension of growth and maturation is seen in the

vertical accumulation of phyterals (plant debris). This body
mirrors the attenuation seen at the surface and it serves as the
substrate to support the changing vegetational communities at the
upper surface. These changes in community character and structure
are reflected in the nature of the sediments derived from their
respective communities. Analysis of these sub-fossil organic
deposits by peat petrology shows that these tree island sites
have been occupied by a succession of communities and environments over the past 5000 years.

Thus employed, peat petrology serves as a useful, new, and
unique paleoecological and paleoenvironmental analytic tool and
has clearly demonstrated paleoecological and paleoenvironmental
changes in southern Florida Holocene sediments.

LITERATURE CITED

Gabe, M. 1976. Histological Techniques. Springer-Verlag,
    New York, Berlin, 1106 p.

International Committee for Coal Petrology. 1963. Inter-
    national Handbook of Coal Petrology, second edition.
    Centre National de la Recherche Scientifique, Paris,
    (unpaginated).

Johansen. 1940. Plant Microtechnique. McGraw-Hill Book Co.,
    New York, 523 p.

Long, R. W. and Lakela, O. 1971. A Flora of Tropical Florida.
    University of Miami Press, Coral Gables, Florida, 962 p.

Radforth, N. W. and Eydt, H. R. 1958. Botanical derivatives
    contributing to the structure of the major peat types.
    Can. Jour. Bot. 36: 153-163.

Smith, W. G. 1968. Sedimentary environments and environmental
    change in the peat-forming area of south Florida. Unpub-
    lished Ph.D. thesis, The Pennsylvania State University,
    254 p.

A LATE- AND POSTGLACIAL POLLEN RECORD FROM CHIPPEWA BOG, LAPEER CO., MI: FURTHER EXAMINATION OF WHITE PINE AND BEECH IMMIGRATION INTO THE CENTRAL GREAT LAKES REGION

Robert E. Bailey and Phyllis J. Ahearn

Central Michigan University

Mt. Pleasant, Michigan 48859

INTRODUCTION

Numerous pollen studies of lake and bog sediments over the past several decades allow for a better understanding of vegetational development in the northeastern portion of North America since the last glaciation. In that many of these also provide a good chronostratigraphy, a reliable data base for establishing patterns of immigration of various forest species during the Holocene is available. Using the available carbon-dated pollen profiles several workers have considered various aspects of immigration patterns. Benninghoff (1964) addressed the possibility of an early Holocene prairie peninsula acting as a filter barrier to the immigration of beech and hemlock into the south-central Great Lakes region. Wright (1968b) considered the immigration of white pine from full-glacial refugia into the Minnesota region, and more recently Jacobson (1979) documented in some detail the immigration pattern for white pines in Minnesota. On a broader scale, Davis (1976) and Bernabo and Webb (1977) presented immigration patterns for various forest species for a good portion of northeastern North America.

Recently, Kapp (1977) considered the immigration phenomena of beech and hemlock in the Great Lakes region and suggested the thumb area of Michigan's lower peninsula as an entry point for beech into this region. In this respect, it is appropriate to consider the pollen record of a strategically located site, Chippewa Bog, as a focal point in addressing questions of white pine and beech immigration into the central Great Lakes region.

53

SITE DESCRIPTION

Chippewa Bog is located within an oak-hickory forest complex
in the eastern half of Michigan's lower peninsula (see Figure 1).
The site is a kettle basin formed at an elevation of ca. 270 m
(890 ft) within a complex interlobate moraine area 5 km north of
Lapeer (43° 07'26" N, 83° 14'28" W) in Lapeer Co., MI., and was
likely deglaciated some time after 13,800 YBP (Farrand and Eschman,
1974). Presently it is part of the Kresge Environmental Education
Center of Eastern Michigan University and locally surrounded by
an upland forest of mixed beech-maple and oak-hickory vegetation.
The bog surface is a closed mat of sphagnum (ca. 222 m x 145 m)
ringed by a water moat at the margin. Vaccinium and Chamaedaphne
occur in abundance in open areas with scattered occurrences of
Larix, Pinus strobus seedlings and Salix sp.

FIELD AND LABORATORY PROCEDURE

Sediment samples were obtained from near the center of the
bog mat in June, 1977 in successive one meter drives using a 5-cm
diameter modified Livingstone sampler (Cushing and Wright, 1965).
7.77 m of sediment was collected, beginning 1.88 m below the
loosely packed sphagnum mat and water layer. Individual cores
were wrapped in Saran Wrap and aluminum foil and transported to
the laboratory and stored under refrigeration. Sediment
characteristics were recorded (see symbolic representation to
the left of Figure 2) prior to selection of unit volume (usually
1.0 cm^3) subsamples for pollen analysis. In order to estimate
pollen concentration a known amount of Eucalyptus pollen was
added to each subsample (Benninghoff, 1962) and processed
following the procedures outlined in Faegri and Iversen (1966).
The pollen-rich residue was suspended in tertiary butyl alcohol
and mounted in silicone oil (12,500 centistokes) on microscope
slides.

A minimum of 200 pollen types were tabulated for each
stratigraphic level and coded on computer cards by pollen type
identification number and abundance per level. These raw data
are available from the authors upon request. In addition, a
minimum of 50 pine grains were measured at selected stratigraphic
levels and coded as white pine, non-white pine or unknown pine
pollen type based upon sculpture pattern (verrucae) on the
underside of the body (Uneo, 1958). The percentage of white
pine is presented as part of the Pinus percentage profile in
Figure 2.

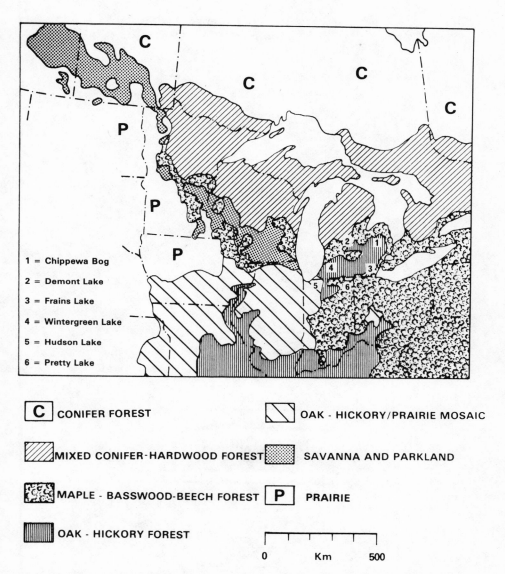

1 = Chippewa Bog

2 = Demont Lake

3 = Frains Lake

4 = Wintergreen Lake

5 = Hudson Lake

6 = Pretty Lake

C CONIFER FOREST

MIXED CONIFER-HARDWOOD FOREST

MAPLE - BASSWOOD-BEECH FOREST

OAK - HICKORY FOREST

OAK - HICKORY/PRAIRIE MOSAIC

SAVANNA AND PARKLAND

P PRAIRIE

0   Km   500

Figure 1. Vegetation map of the central Great Lakes region illustrating locations of six sites addressed in text. (redrawn after Cushing 1965) (1) Ahearn and Bailey 1980, (2) Ahearn 1976, (3) Kerfoot 1974, (4) Manny et al., 1978, (5) Bailey 1972, (6) Williams 1974.

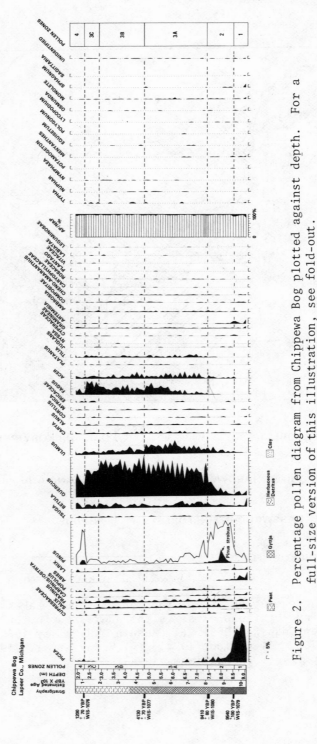

Figure 2.   Percentage pollen diagram from Chippewa Bog plotted against depth.   For a full-size version of this illustration, see fold-out.

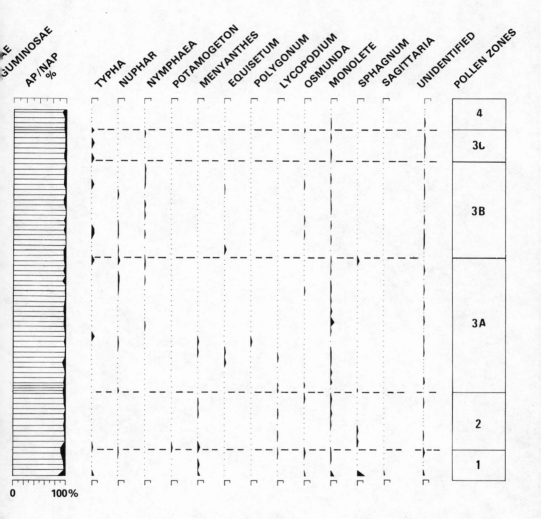

h.

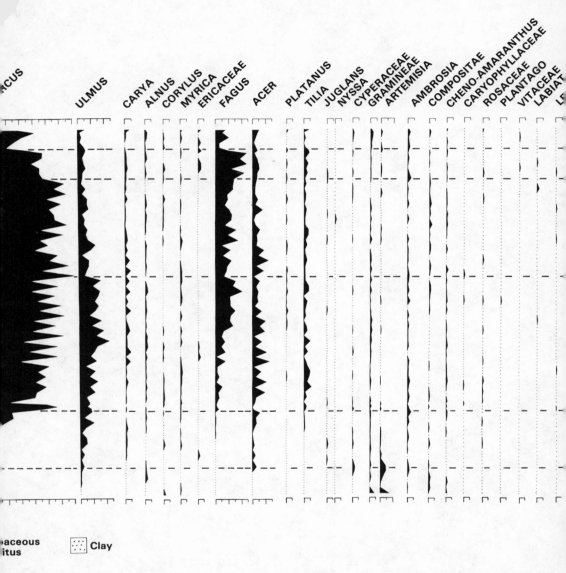

RCUS ULMUS CARYA ALNUS CORYLUS MYRICA ERICACEAE FAGUS ACER PLATANUS TILIA JUGLANS NYSSA CYPERACEAE GRAMINEAE ARTEMISIA AMBROSIA COMPOSITAE CHENO-AMARANTHUS CARYOPHYLLACEAE ROSACEAE PLANTAGO VITACEAE LABIAT

aceous
itus   :::: Clay

re 2.   Percentage pollen diagram from Chippewa Bog plotted against dep

Chippewa Bog
Lapeer Co., Michigan

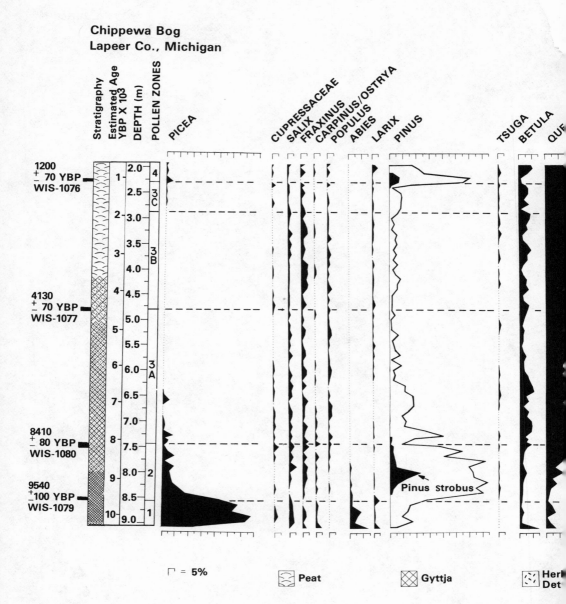

Figu

□ = 5%   ▧ Peat   ⊠ Gyttja   ▨ Herb Det

THE POLLEN DATA

The pollen data for Chippewa Bog, presented graphically in Figure 2 is a percentage pollen diagram wherein the percentage values of 48 pollen types encountered in the study are plotted against depth (from the mat surface) for each of the 74 stratigraphic levels investigated. Sample locations are indicated by horizontal lines in the AP/NAP portion of the diagram. The pollen sum is based upon total pollen, excluding aquatic pollen types and cryptogamic spores. Estimated age is determined from a least squares regression (see Figure 3) of four radiocarbon dates from the stratigraphic levels indicated to the left of Figure 2. A sedimentation rate (SR) of 0.0735 cm/yr was used to estimate pollen influx values. Figure 4 illustrates total AP plus NAP pollen concentration and influx values for each stratigraphic level investigated. In that a constant sedimentation rate was used, the pattern of stratigraphic changes in influx values for specific taxa would generally be reflected in the percentage diagram.

In order to facilitate discussion and interpretation of the pollen stratigraphy four pollen assemblage zones are designated based upon percentage change of the more abundant pollen types. These zones are specific to Chippewa Bog and are not intended to reflect regional pollen stratigraphy, nor imply, by themselves, specific ecological significance (Cushing 1964, West 1970).

Zone 1 (9.06 - 8.56 m, 10280 - 9600 YBP)

This zone is designated as a spruce-fir-herb assemblage where Picea (67.5%), Abies (13.6%), Larix (4.2%) and Betula (14.4%) reach maximum values. Maximum values (12.7%) of non-arboreal pollen types, largely Gramineae and Artemisia, are also reached in this zone. Influx values range from a low of 4.4 x 10^3 to 17.3 x 10^3 grains/cm^2/yr. This assemblage is not unlike that of younger spruce assemblages represented at other sites in southern Michigan and northern Indiana (see Figure 5).

A modern boreal forest analog similar to that in the area of James Bay, Canada is suggested for this assemblage. This is based upon a product-moment correlation value (r) equal to or greater than .95 between the mean pollen spectra of this assemblage and an array of modern pollen surface sample data from eastern North America (Davis and Webb, 1975; Webb and McAndrews, 1976). At this latitude this fossil assemblage appears to represent a transition from an open spruce parkland to a closed spruce forest coincident with a continued warming trend ca. 11,000 to 10,000 YBP.

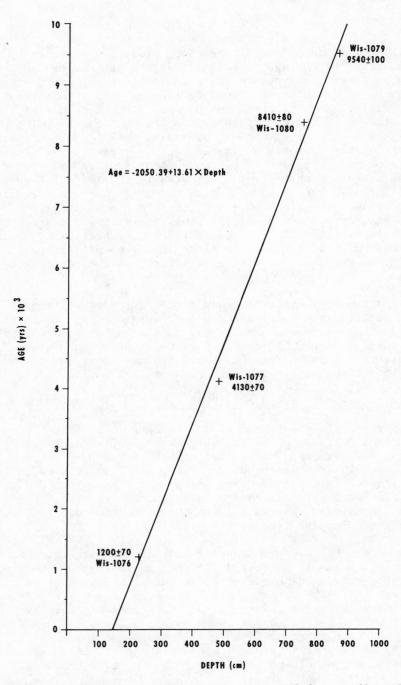

Figure 3. Plot of least square regression of four radiocarbon dates illustrated to left of Figure 2.

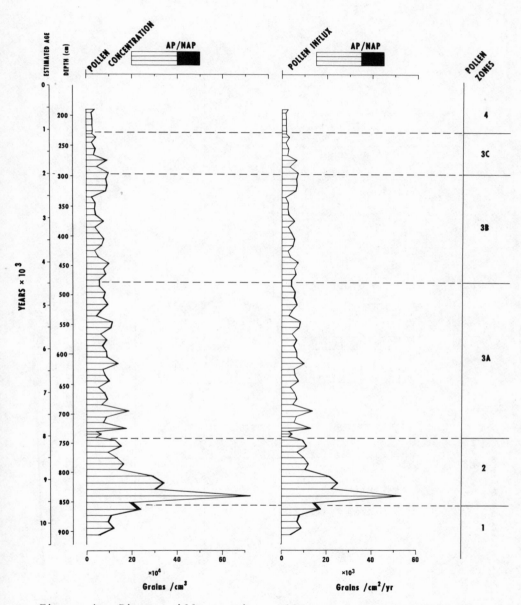

Figure 4. Diagram illustrating pollen concentration and pollen influx plotted against estimated age for Chippewa Bog.

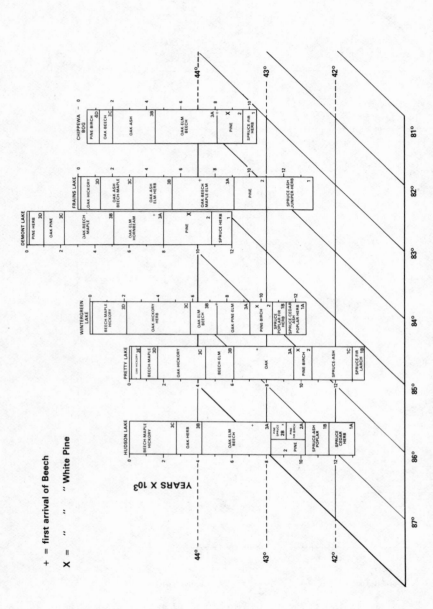

Figure 5. A time-stratigraphic characterization of pollen assemblages for six sites in the central Great Lakes region with reasonably good chronostratigraphy. The assemblage zones for each site are plotted against estimated age and are placed on the lat-long grid to illustrate spatial relationships more appropriately. The zone boundaries for Demont Lake (Ahearn 1976) and Frains Lake (Kerfoot 1974) were designated by the present authors in order to provide better comparison between sites.

Zone 2 (8.56 - 7.44 m, 9600 - 8100 YPB)

    This pine pollen assemblage is characterized by an abundance
of jack and/or red pine pollen, an increase in Quercus and Ulmus,
and the first appearance of certain thermophillous elements
(e.g., Carya, Fagus, Acer and Tilia) in the pollen profile.
Maximum influx values of 53.2 x 10^3 grains/cm^2/yr are recorded
in this zone, of which ca. 72% (38.3 x 10^3) is pine pollen.
Surface sample comparisons with mean pollen spectra of this zone
provide an r-value which suggests a modern analog with a vegetation
type in the upper Great Lakes region.

    The spruce dominated vegetation characteristic of zone 1 was
largely replaced by jack and/or red pine within several hundred
years, likely due to a rapid climatic warming (and drying?) at
the beginning of the Holocene. Such a model fits well with the
regional decline of spruce dominated vegetation at sites in the
south-central Great Lakes region between 11,000 and 10,000 YBP
(Ogden, 1967), and somewhat later at more northerly latitudes
(Wright, 1971a,b). The timing of this event at Chippewa Bog
also fits well with the projections of spruce decline presented
by Bernabo and Webb (1977).

    In the southern portion of the Great Lakes region jack and/or
red pine (the primary colonizers) were partially replaced by a
short lived white pine population and mixed hardwoods. In the
northern portion, today characterized by the mixed coniferous-
hardwood forest (Figure 1), white pine became a major component of
the vegetation and remained so until lumbering activities severely
affected the population size in certain areas (i.e., northern
half of Michigan's lower peninsula). The route of immigration
and time of arrival of white pine from full glacial refugia has
been of interest to paleoecologists for some time (Wright 1964,
1968b; Davis 1976). The exact location of the full glacial
refugia is subject to much conjecture in that the species is
absent from late-Wisconsin sediments at sites in the southeastern
U.S. (Watts 1969, 1970; Whitehead 1964). White pine first appears
ca. 12,500 YBP in northwest Virginia (Craig 1969) and supports
earlier speculation that it was restricted to the continental
shelf (Deevey 1949). White pine apparently moved northward in a
number of directions reaching the New England area ca. 9,000 YBP
(Davis 1969) and the southern portion of the central Great Lakes
region ca. 10,000 YBP (see Figure 6).

    Probably the best understood history of white pine immigration
is presented for the Minnesota area (Jacobson 1979), where the
species moved into the state from the east ca. 7200 YBP. The
population was largely restricted to the northeastern portion of

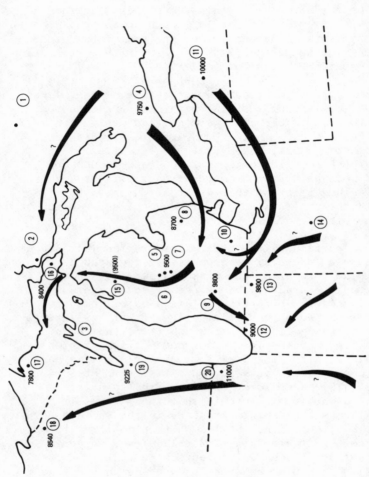

Figure 6. Estimated time of arrival and suggested immigration routes of Pinus strobus in the Great Lakes region. (1) Vincent 1973, (2) Terasmae 1967, (3) Kapp et al., 1969, (4) McAndrews 1973, (5) Ahearn 1976, (6) Gilliam et al., 1967, (7) McMurray et al., 1978, (8) Chippewa Bog, (9) Manny et al., 1978. (10) Kerfoot 1974, (11) Miller 1972, (12) Bailey 1972, (13) Williams 1974, (14) Ogden 1966, (15) Rasmussen and Bailey 1981 (estimated age based on spruce decline of 10,000 YBP, (16) Futyma 1981, (17) Brubaker 1975, (18) Webb 1974, (19) J.C.B. Waddington, unpubl. data), (20) King 1981. For those sites lacking a date, white pine was not separated in the available data.

the state during the mid-postglacial warm period until ca. 4,000
YBP when it began a rapid westward expansion. Based upon the data
in Minnesota and other sites in the Great Lakes region, Ahearn
and Bailey (1980) suggested that white pine may have moved into the
lower peninsula of Michigan from the north. However, given the
recently available data of 8,400 YBP for the arrival of white
pine in the eastern portion of the upper peninsula (site 16,
Figure 6) it is now clear that this could not be the case and that
the immigration of white pine into the lower peninsula of Michigan
was from the south and subsequently moved across the Straits of
Mackinac and then westward. It is equally as plausible, however,
that white pine moved into the upper peninsula from a northerly
route, around the upper portion of Georgian Bay. However, the
lack of white pine separation in pollen profiles at sites in
this area leave the possibility of this route open to question.

The most problematic data addressing the timing of white
pine arrival in the south-central Great Lakes region are from
Volo Bog (site 20, Figure 6) and Hudson Lake (site 12, Figure 6).
At Volo Bog white pine, along with jack and/or red pine were the
major components replacing spruce and may represent white pine
establishment from a source population some place to the south
rather than from the southeast. This hypothesis is strengthened
if the margin of error of the 11,000 year C-14 date for this
event at Volo Bog is minimal.

At Wintergreen Lake and Pretty Lake (sites 9 and 13,
respectively, Figure 6) jack and/or red pine were the primary
colonizers ca. 10,300 YBP, followed by white pine. At Wintergreen
Lake white pine may have been present as early as 10,300, but
only in very small amounts. At Hudson Lake (site 12, Figure 6)
spruce was followed by an assemblage of aspen, fir and ash. Jack
and/or red pine did not appear until about 10,200 YBP; white pine
was not a strong component of the pine assemblage and was not
significantly present until ca. 9,000 YBP. It appears, therefore,
that the source of white pine for sites west and east of the lower
portion of Lake Michigan would not have been the same. It follows
then that the source of white pine for sites in east-central
Wisconsin (site 19, Figure 6) and northeast Wisconsin (site 18,
Figure 6) were from source populations in the Volo Bog area.

At Chippewa Bog, white pine pollen represents a small portion
of the total pine pollen. It is first recorded at ca. 9,300 YBP
and is not present in amounts significant enough to indicate that
it was a component of the vegetation until about 8700 YBP when it
reached a maximum of 22.7% at 7.96 m. White pine was likely
very short-lived in the vegetation and may have represented a
small population which entered the state from the east across

the Pt. Huron-Sarnia border, or from the southeastern portion
of Michigan.

Zone 3 (7.44 - 2.29 m, 8100 - 1070 YBP)

    This zone is characterized as a mixed hardwood assemblage
which spans the majority of the Holocene. Quercus pollen is the
most dominant pollen type (maximum value of 70.7%) and several
other deciduous elements, such as Acer, Fagus, and Ulmus reach
maximum percentages. Pollen influx decreased from the previous
zone, to about 10×10^3 grains/cm^2/yr, except in the early phases
of zone 3A coincident with high percentage values of jack and/or
red pine pollen (see Figure 4). Three subassemblages are
designated largely based upon changes in percent expression of
Fagus and Ulmus pollen.

    Subzone 3A (7.44 - 4.81 m, 8100 - 4500 YBP) is designated as
an oak-elm-beech subassemblage. With continued climatic warming
jack and/or red pine populations decreased rapidly and were replaced by birch, oak and elm. This successional event is recorded
about 1,000 years earlier in the southern portion of the central
Great Lakes region (see Figure 5) in comparison to similar events
at Chippewa Bog and Demont Lake.

    One of the more interesting features of zone 3A is the
establishment of Fagus at Chippewa Bog at 8100 YBP, adding considerable support to Kapp's (1977) hypothesis that beech entered
the central Great Lakes region from the east via Ontario, probably
across the Pt. Huron-Sarnia border (see Figure 7). Fagus may
have moved into southern Ontario from the south across the Lake
Erie basin during low water stages, or from the eastern portion
of Ontario north of Lake Ontario. Given the distribution of
dates for Fagus establishment in northern Indiana and southern
Michigan it is difficult to ascertain whether the source population for these localities was central Ohio or the thumb area of
Michigan. Fagus could have moved slowly in a northwesterly
direction from central Ohio (site 14) and northeastern Ohio
(Shane 1975) where it was present as early as ca. 10,000 YBP. It
seems more likely, however, that the early Holocene Hypsithermal,
registered as nearby as central and northeastern Illinois at ca.
8,000 YBP (King 1981), created sufficient moisture stress in
northern Indiana to preclude rapid movement from central Ohio
(Benninghoff 1964; Wright 1968a, 1976). Support for an extension
of moisture stress conditions into the central Great Lakes region
between 8,000 and 6,000 YBP is seen at Chippewa Bog where oak,
birch, elm and jack and/or red pine, and to some extent basswood
and maple, were favored. Similar patterns are observed at other

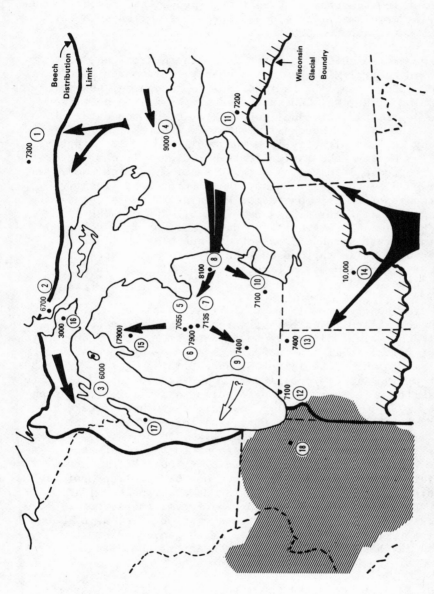

Figure 7. Estimated time of arrival of *Fagus grandifolia* and suggested immigration routes into the central Great Lakes region. Map modified after Kapp (1977). Shaded area indicates prairie vegetation. All sites are as indicated in Figure 6, except site 18, which is the location of McGinnis Slough (Bailey, unpub. data).

sites in the central Great Lakes region (see Figure 5). At
Chippewa Bog jack and/or red pine, oak and probably birch were
favored in the uplands, while elm, basswood and maple were favored
in the moist lowland areas. Even though beech was present, population sizes were small and did not become established in large
quantities until about 6,000 YBP.

Immigration of beech to sites in southern Michigan, northern
Indiana, and central Michigan was probably from source points in
the Chippewa Bog area. It is important to recognize, however,
that the time of arrival and suggested routes are to be viewed
with some degree of individual interpretation of the data. In
this case, the time of beech arrival at Frains Lake (site 10) is
reported by reading directly from the published pollen diagram;
the data for Pretty Lake (site 13) is based on a linear regression
of the corrected C-14 dates plus the data of land clearance. On
the other hand, Webb (pers. comm.) has interpolated linearly
between two reported C-14 dates for Fagus at 2% and suggests a
date of ca 8,000 YBP for Frains Lake and 6,700-6,400 YBP for
Pretty Lake. The Wintergreen Lake data (site 9) is based upon
a curvilinear regression of corrected dates plus the land
clearance date, while the Hudson Lake date is based upon a
curvilinear regression of uncorrected dates plus the land
clearance date.

The cluster of sites in central Michigan (sites 5,6,7)
provide some difficulty in establishing a date for beech immigration to this area, although a mean date of 7,360 YBP would
place its arrival in line with projections made by Bernabo and
Webb (1977). At Cub Lake (site 15) an estimated age of 7,900 YBP
for beech arrival is extrapolated from an assumed 10,000 YBP date
for the shift in the spruce to pine pollen assemblage (Rasmussen
and Bailey 1981). It is likely, however, that the Cub Lake date
is too old in that Webb (pers. comm.) estimates beech arrival at
Green Lake, 20 miles northwest of Cub Lake, to be ca. 5,400 YBP.
This estimate is based on some adjustment to the C-14 dates
reported by Lawrenz (1975) and a date of 4,800 YBP for the
hemlock fall (see Davis 1977) in the Green Lake pollen profile,
and is more in line with the 6,000 YBP date for Beaver Island
(site 3). A 3,000 YBP date provided by Futyma (pers. comm.) for
three localities (site 16) indicates a considerable lag of beech
immigration to the upper peninsula of Michigan. The source
population may have been from the south across the Straits, or
north of Georgian Bay.

Between ca. 6,000 and 4,000 YBP mesic conditions increased
in the central Great Lakes region. Fagus and Ulmus are commonly
represented in the pollen data at the 20% level, and apparently

were well established as major components of the vegetation (see
Figure 5). This feature is evident at Chippewa Bog where jack
and/or red pine and birch are replaced by these mesic species.
It is possible that during this time period beech expanded
regionally to some areas outside of its present distribution
limit (e.g., site 1). It was clearly established at Seidel Lake
(site 17) during a good portion of the Holocene, but the exact
arrival time is difficult to determine from data presented by
West (1961). Several grains (ca. 0.8%) are reported as early as
8,600 YBP from a more recent analysis of this site (unpubl. data,
J.C.B. Waddington; Webb, pers. comm.) and may represent only
windblown grains. Fagus is represented sporadically at ca. 2.5%
through the Holocene beginning as early as 6,000 YBP at Chatsworth
Bog in central Illinois and Volo Bog in the extreme northeastern
section of Illinois, but is probably due to long distance
transport (King 1981). At McGinnis Slough (site 18) Fagus occurs
at ca. 14% in the upper several meters of sediment (Bailey, unpubl.
data) and consistently at ca. 10% beginning ca. 6,000 YBP at Cedar
Lake in the extreme northwest section of Indiana (Bailey 1979).
From these data it appears that at some time during the Holocene,
beech did extend beyond its present distribution limits.

Zone 3B (4.81 - 2.91 m, 4500 - 1900 YBP) is characterized as
an oak-birch-ash subassemblage. Oak maintains high percentage
values (60% range), birch fluctuates between 2-10%, ash is represented at values greater than 5% in two phases of this zone, and
jack and/or red pine increase somewhat over the upper portion of
zone 3A. Influx values range between 5-7 x 10^5 grains/cm^2/yr and
reach low values of 1.8 - 3.1 x 10^5 grains/cm^2/yr.

One of the major features of this zone is the marked decline
in Ulmus, from 16 to 4% over 10 cm (135 years) at ca. 4,500 YBP.
A similar event occurs in western European pollen diagrams at
about this same time period (Tauber 1965). In the central Great
Lakes region this drop in elm is a consistent feature between
ca. 4,500 and 4,000 YBP and most always occurs with a similar
decrease in beech percentages. At some sites (e.g., Hudson Lake
and Wintergreen Lake, see Figure 5) herbaceous pollen also increases significantly (15-22%). It appears that the elm drop,
when correlated with these other events, is more of a climatic
phenomenon rather than other causal factors such as disease
(Tauber 1965). At Chippewa Bog, beech, elm, birch, ash, and
jack and/or red pine oscillations may be reflective of alternating
periods of water stress over a several thousand year period.

Zone 3C (2.91 - 2.35 m, 1900 - 1150 YBP). In this subzone
beech increases to maximum values (greater than 23%) and jack
and/or red pine increase to about 10% largely at the expense of

oak. Pollen influx, initially about 10×10^3 grains/cm^2/yr drops
to about 5×10^3 grains/cm^2/yr beginning at the middle of this
zone. It appears that beech was well established during this time
period as a major part of the regional vegetation, probably due to
an increase in moisture availability. Water levels in the Chippewa
Bog basin may have been higher at this time and the development
of a bog mat may have begun here as suggested by the presence of
Ericaceae pollen types in the range of 2-3% in the upper part of
this zone.

Zone 4 (2.35 - 1.91 m, 1150 - 550 YBP)

This zone is characterized as a pine-hardwood assemblage
where pine increases markedly, from 15 to 61% over 10 cm (135 yrs),
and spruce, Ambrosia and Gramineae increase slightly while most
all hardwoods decrease synchronously. Fire incidence is the most
likely explanation for this change in the pollen record in that the
first part of the assemblage is dominated by jack and/or red pine
which then drops off as hardwoods increase near the top of the
section. Whether or not this event is coupled with a change in
climatic conditions (i.e., increase in temperature or decrease
in precipitation), which would support fire incidence on a
regional scale, is difficult to determine. It is important to
recognize that evidence for fire in the recent Holocene is present
at other localities.

At Demont Lake (Ahearn 1976) a similar event is reported
where NAP increases (especially Gramineae) together with a major
increase in pine and fluctuations in concentration of charcoal
particles prior to the pine peak. At Lake of the Clouds in
Minnesota (Swain 1973) the pollen record remains virtually un-
changed over the past 1500 years, although varved chronology
provides a well defined charcoal particle profile indicating
several fire events over this same time period.

At Chippewa Bog, no charcoal particles are found in the
sediment. However, jack and/or red pine would likely have been
the primary colonizers after fire in this area. Given the modern
distribution of these species (Voss 1972) a seed source could
have been available. The lack of significant NAP increase
following the pine peak may be attributed to the upland vegetation
and the bog vegetation itself acting as a filter barrier to pollen
from outlying herbaceous communities.

The uppermost section at Chippewa Bog is truncated and does
not record the characteristic NAP increase indicative of
European settlement. The pollen record for the past 500 years
in this area is therefore not available.

SUMMARY

The pollen record for Chippewa Bog begins ca. 10,500 YBP with a basal spruce-fir-herb assemblage characteristic of a boreal forest in the James Bay region of Canada and was succeeded by a jack and/or red pine dominated vegetation ca. 9500 YBP. White pine was not a major component of the pine forest and not present in significant quantities until ca. 8,700 YBP. The source of the Chippewa Bog local white pine population may have been to the east, while the migration of white pine into majority of the central Great Lakes region was likely from the south.

At ca. 8,100 YBP, the pine assemblage was replaced by a mixed hardwood assemblage dominated by oak pollen, with fluctuations of several mesic elements (i.e., elm and beech) throughout much of the Holocene. Beech most probably entered Michigan from the east, across the Pt. Huron-Sarnia border at ca. 8,100 YBP and subsequently migrated south, southwest and to the east over the next 1,000 years. Rapid movement from the thumb area of Michigan and from the central Ohio region may have been restricted by moisture stress conditions in the south-central Great Lakes region between 8,000 and 6,000 YBP. Indeed, several periods of moisture stress during the Holocene may be evident in a number of pollen profiles from the central Great Lakes region, based on the fluctuations in percentages of beech and to some extend elm pollen.

The upper assemblage (Zone 4) of Chippewa Bog likely records a fire of unknown dimension just prior to ca. 1,200 YBP, with reforestation by jack and/or red pine, lasting for ca. 300 years. This pine dominated vegetation was succeeded by an assemblage similar to that before the fire, but with more pine present. The characteristics of the vegetation between ca. 500 YBP and land clearance in the area of Chippewa Bog is not known due to truncation of the upper section of sediment.

LITERATURE CITED

Ahearn, P. J. 1976. Late-glacial and postglacial pollen record from Demont Lake, Isabella County, Michigan. Unpubl. Senior Thesis, Alma College, 17 pp.

Ahearn, P. J. and Bailey, R. E. 1980. Pollen record from Chippewa Bog, Lapeer County, Michigan. Michigan Academician. 12(3): 297-308.

Bailey, R. E.   1972.   Late- and postglacial environmental changes
    in Northwestern Indiana.   Ph.D. thesis, Indiana Univ.
    (unpubl).

-------.   1979.   Patterns of vegetational succession in the
    central Great Lakes region.   Contrib. Pap., Paleoecol.
    Sec., ESA meeting, Stillwater, OK., August, 1979.

Benninghoff, W. S.   1962.   Calculation of pollen and spore
    density in sediments by addition of exotic pollen in known
    quantities.   (Abs) Pollen et Spores 4: 332-333.

-------.   1964.   The Prairie Peninsula as a filter barrier to
    postglacial plant migration.   Indiana Acad. Sci. Proc. 72:
    116-124.

Bernabo, J. C. and Webb, T., III.   1977.   Changing patterns in
    the Holocene pollen record of northeastern North America:
    A mapped summary.   Quaternary Research. 8:   20-53.

Craig, A. J.   1969.   Vegetational history of the Shenandoah
    Valley, Virginia.   GSA Spec. Pap. 123: 283-296.

Cushing, E. J.   1964.   Application of the Code of Stratigraphic
    Nomenclature to Pollen Stratigraphy.   Minn. Geol. Surv.
    (Unpubl. Ms.).

-------.   1965.   Problems in the Quaternary phytogeography of
    the Great Lakes region.   p. 403-416 in: H. E. Wright, Jr.
    and D. G. Frey, (eds.)   The Quaternary of the United
    States.   Princeton Univ. Press, Princeton, N.J.

Cushing, E. J. and Wright, H. E., Jr.   1965.   Hand-operated
    piston cores for lake sediments.   Ecology.   46: 380-384.

Davis, M. B.   1969.   Climatic changes in southern Connecticut
    recorded by pollen deposition at Rogers Lake.   Ecology.
    50: 409-422.

-------.   1976.   Pleistocene Biogeography of Temperate
    Deciduous Forests.   Geoscience and Man.   13: 13-26.

-------.   1977.   Outbreaks of forest pathogens in Quaternary
    history.   Proc. IV Intern. Conf. on Palynology, Lucknow,
    India (pers. comm.).

Davis, R. B. and Webb, T., III. 1975. The contemporary dis-
    tribution of pollen in eastern North America: a comparison
    with the vegetation. Quaternary Research. 5: 395-434.

Deevey, E. S. 1949. Biogeography of the Pleistocene. Part 1,
    Europe and North America. Geol. Soc. Amer. Bull. 60:
    1315-1416.

Farrand, W. R. and Eschman, D. F. 1974. Glaciation of the
    Southern Peninsula of Michigan: A review. Michigan
    Academician. 7: 31-56.

Gaegri, K. and Iverson, J. 1966. Textbook of Pollen Analysis.
    Munksgaard, Denmark. 237 pp.

Futyma, R. P. 1981. Holocene vegetational history of Michigan's
    eastern Upper Peninsula. Paper presented at Mich. Acad.
    Sci., Arts, and Letters, March 20, 1981, Ann Arbor, MI.

Gilliam, J. A., Kapp, R. O. and Bogue, R. D. 1967. A post-
    Wisconsin pollen sequence from Vestaburg bog, Montcalm
    County, Michigan. Pap. Mich. Acad. Sci., Arts, Letters
    52: 3-17.

Jacobson, G. L., Jr. 1979. The paleoecology of white pine
    (Pinus strobus) in Minnesota. Jour. Ecology. 67: 697-726.

Kapp, R. O. 1977. Late Pleistocene and Postglacial Plant
    Communities of the Great Lakes Region. pp. 1-27. In:
    R. C. Romans (ed.) Geobotany. Plenum Publ. Corp.

Kerfoot, W. C. 1974. Net accumulation rates and the history
    of Cladoceran communities. Ecology. 55: 51-61.

King, J. E. 1981. Late Quaternary vegetational history of
    Illinois. Ecol. Monogr. 51: 43-62.

Lawrenz, R. 1975. Biostratigraphic study of Green Lake,
    Michigan. M.S. thesis, Central Michigan Univ. (unpubl.).

Manny, B. A., Wetzel, R. G., and Bailey, R. E. 1978.
    Paleolimnological sedimentation of organic carbon, nitrogen,
    phosphorus, fossil pigments, pollen, and diatoms in a
    hypereutrophic, hardwater lake: A case history of
    eutrophication. Pol. Arch. Hydrobiol. 25: 243-267.

McAndrews, J. H.   1973.   Pollen analysis of the sediments of
    the Great Lakes of North America, p. 76-80.   In: Palynology,
    Holocene, and Marine Palynology, Proc. III Int. Paly,
    Conf. Publishing House Nauka, Moscow.

McMurray, M., Kloos, G., Kapp, R., and Sullivan, K.   1978.
    Paleoecology of Crystal Marsh, Montcalm County, based on
    Macrofossil and Pollen Analysis.   Michigan Academician.
    10: 403-417.

Miller, N. G.   1973.   Late-glacial and postglacial vegetation
    change in southwestern New York State.   New York State
    Mus. and Sci. Serv., Bull. 420, 102 p.

Ogden, J. G., III.   1966.   Forest history of Ohio.   Radiocarbon
    dates and pollen stratigraphy of Silver Lake, Logan
    County, Ohio.   Ohio J. Sci. 66: 387-400.

--------.   1967.   Radiocarbon and pollen evidence for a sudden
    change in climate in the Great Lakes region approximately
    10,000 years ago, p. 117-127.   In: Cushing, E. J. and
    Wright, H. E., Jr. (eds.), Quaternary Paleoecology.
    Yale Univ. Press.

Rasmussen, J. B. and Bailey, R. E.   1981.   Pollen record from
    Cub Lake, Kalkaska County, Michigan.   Paper presented at
    Mich. Acad. Sci., Arts, and Letters, March 20, 1981,
    Ann Arbor, MI.

Shane, L. C.   1975.   Palynology and radiocarbon chronology of
    Battaglia Bog, Portage County, Ohio.   Ohio Jour. Sci. 75:
    96-102.

Swain, A. M.   1973.   A history of fire and vegetation in
    northeastern Minnesota as recorded in lake sediments.
    Quaternary Research.   3: 383-396.

Terasmae, J.   1967.   Postglacial chronology and forest history
    in the northern Lake Huron and Lake Superior regions.
    p. 45-58, In: Cushing, E. J. and Wright, H. E., Jr.,
    (eds.), Quaternary Paleoecology.   Yale Univ. Press.

Tauber, H.   1965.   Differential pollen dispersion and the
    interpretation of pollen diagrams.   Danmarks Geol.
    Unersgelse, II, 89: 1-69.

Ueno, J.   1958.   Some palynological observations of Pinaceae.
    J. Inst. Polytechnics, Osaka City Univ., Ser. D 9: 163-186.

Watts, W. A. 1969. A pollen diagram from Mud Lake, Marion
    County, north-central Florida. Geol. Soc. Amer. Bull.
    90: 631-642.

--------. 1970. The full-glacial vegetation of northwestern
    Georgia. Ecology. 51: 17-33.

Webb, T., III. 1974. A vegetational history from northern
    Wisconsin: evidence from modern and fossil pollen. Amer.
    Midl. Nat. 92: 12-34.

Webb, T., III and McAndrews, J. H. 1976. Corresponding patterns
    of contemporary pollen and vegetation in central North
    America. Geol. Soc. Amer. Mem. 145: 267-302.

West, R. G. 1961. Late- and postglacial vegetation history in
    Wisconsin, particularly changes associated with the Valders
    readvance. Amer. Jour. Sci. 259: 766-783.

--------. 1970. Pollen zones in the Pleistocene of Great
    Britain and their correlation. New Phytol. 69: 1179-1183.

Whitehead, D. R. 1964. Fossil pine pollen and full glacial
    vegetation in southeastern North Carolina. Ecology. 45:
    767-777.

Williams, A. S. 1974. Late-glacial-Postglacial vegetational
    history of the Pretty Lake region, northeastern Indiana:
    Hydrologic and biological studies of Pretty Lake, Indiana.
    U.S.G.S. Prof. Pap. No. 686. Wash., D.C.

Wright, H. E., Jr. 1964. Aspects of the early postglacial
    forest succession in the Great Lakes region. Ecology. 45:
    439-448.

--------. 1968a. History of the prairie peninsula. p. 78-88.
    In: Bergstrom, R. E. (ed.), The Quaternary of Illinois.
    Univ. Illinois. College Agric. Spec. Publ. 14.

--------. 1968b. The roles of pine and spruce in the forest
    history of Minnesota and adjacent areas. Ecology. 49:
    937-955.

--------. 1971a. Late Quaternary vegetational history of North
    America. p. 425-464, In: Turekian, K. (ed.), The Late
    Cenozoic Glacial Ages. Yale Univ. Press.

Wright, H. E., Jr. 1971b. Retreat of the Laurentide Ice Sheet
    from 14,000 to 9,000 years ago. Quaternary Research. 1:
    316-330.

-------. 1976. The dynamic nature of Holocene vegetation. A
    problem in paleoclimatology, biogeography, and stratigraphic
    nomenclature. Quaternary Research. 6: 581-596.

LATE-WISCONSIN DEGLACIATION AND MIGRATION OF SPRUCE INTO

SOUTHERN ONTARIO, CANADA

J. Terasmae

Brock University

St. Catharines, Ontario L2S 3A1

ABSTRACT

Palynological studies of about 10 lake sediment cores, paleobotanical data, and recent geological investigations of deglaciation and the Great Lakes history of southern Ontario (supported by about 50 radiocarbon dates) indicate that spruce and dwarf-shrub tundra migrated into southern Ontario during two different episodes in Late-Wisconsin time. First, about 13,500 years BP during the Mackinaw Interstadial that coincided with a low-water phase in Lake Erie basin it was possible for spruce to migrate into southern Ontario across Lake Erie basin to establish founder populations. Spruce probably survived the ice readvances during the following Port Huron Stadial (about 13,000 years BP) in southern Ontario. These founder populations developed rapidly into spruce forest about 12,500 years BP during the Two Creeks Interstadial (as indicated by palynological data) when the second wave of migration of spruce occurred across Lake Erie basin during the low-level Early Lake Erie phase (beginning about 12,600 years BP) that followed the high-level glacial Lake Whittlesey and Lake Warren phases.

Spruce cones at Brampton near Toronto have been radiocarbon dated at 12,320 ± 360 years BP and a radiocarbon age of about 12,500 years BP is proposed for the lower boundary of the spruce pollen zone in southern Ontario that was preceded by a herb pollen zone that represents a dwarf-shrub tundra in which some spruce populations were present locally to account for the 30 - 50 % of spruce pollen found in this zone.

The Late-Wisconsin history of spruce and deglaciation in southwestern Ontario appears to be related to the presence of mastodon in this region and provides a partial answer to its extinction.

INTRODUCTION

During the last 10 years there has been a resurgence of studies relating to the Late-Wisconsin deglaciation history of southern Ontario. The mapping program of surficial deposits by the Ontario Geological Survey has been one of the most significant factors in these investigations that have resulted in a number of revisions of the previously proposed chronological sequences of late-glacial events in southwestern Ontario. For example, the assignment of Port Bruce Stadial age to the Paris and Galt Moraines (previously believed to be of Port Huron Stadial age; about 13,000 years BP) has implied earlier deglaciation of the central part of southwestern Ontario, and has required some revision of the chronology of glacial lake phases in the Lake Erie basin.

Both the revised history of deglaciation and the sequence of glacial lake phases, that depended on the retreat of ice, are closely linked to the progress of Late-Wisconsin revegetation in southern Ontario.

The purpose of this report is to investigate the Late-Wisconsin history of vegetation (especially spruce) in southwestern Ontario with references to the current understanding of the deglaciation chronology.

GLACIAL HISTORY

A comprehensive summary of the glacial and postglacial history of Great Lakes region, especially covering the last 14,000 years, was published by Prest (1970) and it contained a number of revisions of earlier reviews of deglaciation of southern Ontario (Hough, 1958, 1963; Chapman and Putnam, 1966). Deglaciation histories concerned more specifically with southwestern Ontario (including the Lake Erie basin) were published by Lewis et al. (1966), Dreimanis (1969), Lewis (1969), Sly and Lewis (1972), Terasmae et al. (1972), and Dreimanis and Goldthwait (1973).

Dreimanis and Karrow (1972) proposed a new classification of the Wisconsinan Stage for southern Ontario, and further revisions of the deglaciation history of southwestern Ontario were proposed by Cowan et al. (1975), Karrow et al. (1975), Dreimanis (1977), Cowan et al. (1978), and Barnett (1979).

Many other reports, related mostly to the mapping program of surficial deposits by the Ontario Geological Survey, were published in the 1970-s and covered glacial history of specific map-areas in southern Ontario.

The deglaciation of southwestern Ontario was controlled essentially by the interaction of four major marginal lobes of the Late-Wisconsin ice sheet (Cowan et al., 1975); the Lake Huron lobe, the Lake Erie lobe, the Lake Ontario lobe, and the Georgian Bay lobe.

All of southwestern Ontario was covered by ice during the Nissouri Stadial (between 24,000 and 16,000 years BP). A significant ice retreat occurred during the Erie Interstadial (about 16,000 to 15,500 years BP; Mörner and Dreimanis, 1973) when most of southwestern Ontario was probably deglaciated (Dreimanis, 1977), and the readvance of ice during the Port Bruce Stadial extended again south of Lake Erie (Dreimanis and Goldthwait, 1973) when the Port Stanley Till was deposited in southwestern Ontario.

The mass budget of the ice sheet must have been negative during the Port Bruce Stadial because some of southwestern Ontario was deglaciated at that time as indicated by the end moraines of Port Bruce Stadial age (Fig. 1) in southwestern Ontario (Cowan et al., 1975).

It is interesting to note that there is a possibility that some relatively small parts of central southwestern Ontario may not have been covered by ice during the Port Bruce Stadial. For example, Cowan et al. (1975) show some areas of Catfish Creek Till (Nissouri Stadial age) exposed at the surface south of Waterloo and Woodstock. Radiocarbon dates on oldest organic sediments in Sunfish Lake (15,950 ± 850 years BP, I-4652; Sreenivasa, 1973) and Lake Hunger (15,180 ± 200 years BP, BGS-266; Winn, 1977) in this area have been discredited as being too old on account of contamination by "old carbon" or the "hardwater effect", or possibly because these dates did not fit the accepted deglaciation chronology of southwestern Ontario. Perhaps this matter deserves to be reconsidered in view of the probability that these dates are not too old by 3000 or 4000 years as suggested and, in fact, it is the deglaciation history that should be revised.

The central part of southwestern Ontario was definitely deglaciated during the Port Bruce Stadial when the four ice lobes separated and deposited a ring of end moraines (Fig. 1) around

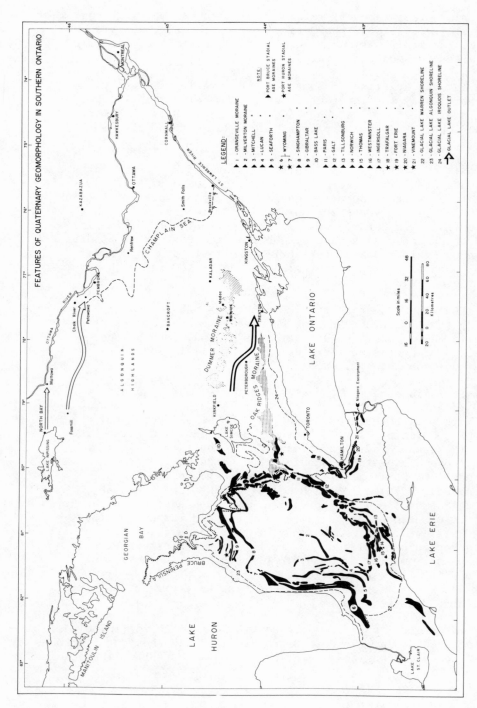

Fig. 1. Map of Late-Wisconsin end moraines and other geomorphological features of southwestern Ontario.

the "Ontario Island" (the name used for the earliest deglaciated part of central southwestern Ontario by Chapman and Putnam, 1966).

A sequence of ice-dammed glacial lakes in the Lake Erie basin was closely related to the progress of deglaciation, and the levels of these lakes fluctuated in response to the opening and closing of the different outlets as well as the beginning of the differential crustal rebound in Late-Wisconsin time (Fig. 2).

It is reasonable to assume that glacial lakes developed in the Lake Erie basin during the Erie Interstadial, including a postulated low-water phase (Lake Leverett) as indicated in Fig. 2. However, most evidence of these lakes was obliterated by ice advance during the subsequent Port Bruce Stadial.

The stepwise retreat of the Lake Erie ice lobe is indicated by three end moraines that define the western, central, and eastern basins in Lake Erie (Sly and Lewis, 1972). Glacial Lake Maumee developed in the western part of Lake Erie basin during the Port Bruce Stadial, and a low-water phase (Lake Ypsilanti) has been postulated to have existed in the Erie basin prior to deposition of the Wentworth Till and the glacial Lake Arkona phase (Fig. 2). Barnett (1979) proposed that the Wentworth Till was deposited during the Port Bruce Stadial (rather than the Port Huron Stadial as postulated previously), and that a low-water phase (the "Simcoe Delta" phase) existed in the Lake Erie basin during the Mackinaw Interstadial and prior to deposition of the Halton Till (Port Huron Stadial). Barnett's study implies that glacial Lake Whittlesey is younger than the Paris, Galt, and Tillsonburg Moraines (Fig. 1) that were previously believed to belong in the Port Huron Stadial. This revision allows more time for development of glacial lakes Whittlesey, Warren, and other late phases (Fig. 2) prior to the low-water Early Lake Erie phase that began about 12,600 years BP (Lewis, 1969) and preceded the glacial Lake Iroquois phase in the Lake Ontario basin about 12,500 years BP.

In terms of the present study, probably the most significant implication of the revisions of deglaciation history of southern Ontario, as supported by recent studies, is the existence of low-water levels in the Lake Erie basin prior to the Early Lake Erie phase, and the possibility that at least some parts of central southwestern Ontario may not have been covered by ice during most of the Port Bruce Stadial or even since the Erie Interstadial. This statement does not exclude the presence of ice south of Lake Erie during the Port Bruce Stadial when the Lake Erie lobe readvanced to end moraine positions of that age both south and north of the Lake Erie basin.

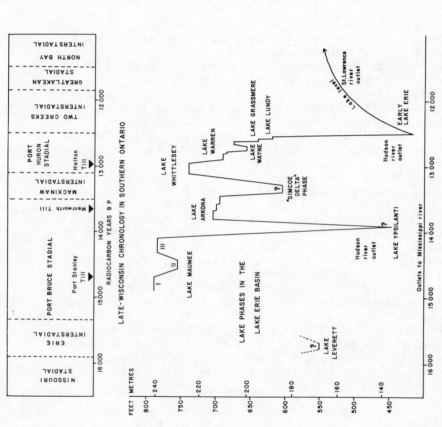

Fig. 2. Late-Wisconsin sequence of glacial lake phases and lake level changes in the Lake Erie basin, and the proposed chronology of stadial and interstadial intervals.

PALYNOLOGY AND PALEOBOTANY

In the early 1960-s the writer compiled a pollen diagram
(published in Karrow, 1963) from the Crieff kettle bog in the
Galt Moraine and a radiocarbon date of 11,950 ± 350 years BP
(I:GSC-29) was obtained for the oldest organic sediment (algal
gyttja). The pollen record extended at least 30 - 50 cm deeper
in the silty clay beneath the dated gyttja and was dominated by
spruce pollen (60 - 70 %). If one accepts the calculated sedimentation rate of about 0.015 cm yr^{-1}, proposed by Mott and
Farley-Gill (1978) for similar sediments (dated at 12,400 ± 180
years BP, GSC-1156) in Maplehurst Lake in the same general area,
then the pollen record in Crieff kettle bog probably extends to
at least 13,000 years BP. This reasoning supports the revised
age of the Galt Moraine (Port Bruce Stadial) and implies that
spruce was present in the central part of southwestern Ontario
during the Mackinaw Interstadial (about 13,500 years BP),
especially if one considers that probably several hundred years
were required for the melt-out of buried ice that formed the
Crieff kettle which, in turn, may explain the absence of a herb
pollen zone beneath the spruce pollen zone in this bog.

Although this interpretation was not possible in the early
1960-s on the basis of only one pollen diagram and a single
radiocarbon date, it now appears reasonable in view of studies
made in the writer's laboratory during the last 10 years, including about 10 pollen diagrams from lake sediment cores,
palynological studies of Late-Wisconsin age deposits at 15 sites,
and about 50 radiocarbon dates from southwestern Ontario. In
addition, other studies for example by Anderson (1971),
Sreenivasa (1973), McAndrews et al. (1974), Karrow et al. (1975),
Mott and Farley-Gill (1978) in southwestern Ontario contain
palynological data that are in agreement with those obtained by
the writer.

A generalized palynostratigraphic sequence for southwestern
Ontario is presented in Fig. 3. The present study is concerned
primarily with the earliest two pollen zones of that sequence -
the herb pollen zone and the spruce pollen zone.

As proposed by Karrow et al. (1975) the upper boundary of
the spruce pollen zone has a radiocarbon age of about 10,600
years BP in southwestern Ontario. Mott and Farley-Gill (1978)
obtained a radiocarbon date of 12,400 ± 180 years BP (GSC-1156)
for the lower boundary of the spruce pollen zone. This date was
corrected to 12,500 ± 180 years BP (δ ^{13}C=-21.0 ^{o}/oo), and they
proposed an "adjusted age" of 12,020 ± 180 years BP for this
sample on the basis of an estimated age of 480 years for the

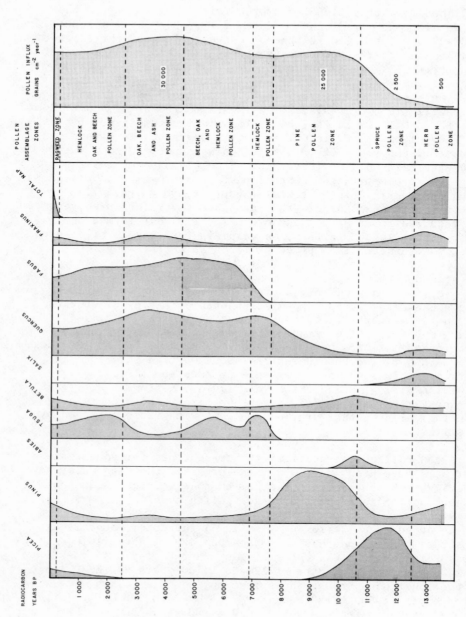

Fig. 3. Late-Wisconsin and postglacial palynostratigraphy and radiocarbon chronology of southwestern Ontario.

sediment surface. I believe that this adjustment may be open for question.

A recent study by the writer (Terasmae and Matthews, 1980) yielded a radiocarbon date of 12,320 ± 360 years BP (BGS-551) for a sample of white spruce cones at Brampton, Ontario, and indicated that spruce was growing in the Toronto area by about 12,500 radiocarbon years BP. Samples of spruce wood from southwestern Ontario (Winn, 1977) have been dated at 11,660 ± 140 years BP (BGS-407) and 12,130 ± 150 years BP (BGS-375).

The lower boundary of the spruce pollen zone has been deter- mined by both pollen percentage and influx calculations (Mott and Farley-Gill, 1978). The influx and concentration of pollen decrease markedly in the spruce pollen zone, and the proportion (percentage) of herb pollen increases. These palynostratigraphic changes have been used to define the lower boundary of the spruce pollen zone that has a radiocarbon age of about 12,500 years BP in southwestern Ontario.

In terms of vegetation, spruce forest became established in southwestern Ontario during the spruce pollen zone, that was preceded by an essentially treeless vegetation during the herb pollen zone. However, the percentage of spruce pollen appears to remain rather high (30 - 40%, Mott and Farley-Gill, 1978; and at least 50 - 60 %, Winn, 1977) in the herb pollen zone, and this has been interpreted to indicate that some spruce was probably growing in central southwestern Ontario about 13,000 years ago. Three radiocarbon dates in this range were obtained by Winn (1977); 12,950 ± 220 years BP (BGS-421), Cornell bog; 12,930 ± 400 years BP (BGS-366), Colles Lake; 13,230 ± 210 years BP (BGS-234a), Lake Hunger. In all these cases the pollen record extends beyond the dated oldest organic lake sediments and there is a reasonable possibility that spruce could have been growing in southwestern Ontario well before 13,000 years ago in Late- Wisconsin time.

It is probable that the vegetation represented by the herb pollen zone (characterized by pollen of Cyperaceae, Gramineae, Artemisia, Shepherdia canadensis, Compositae, and various other species indicating treeless vegetation) resembled dwarf-shrub tundra with spruce present at ecologically most favourable sites. The discovery of Dryas integrifolia leaves at Brampton (Terasmae and Matthews, 1980) supports this interpretation.

The presence of Late-Wisconsin fossil ice-wedge polygons near Kitchener (Morgan, 1972) also supports the postulated tundra-like environmental conditions in central southwestern Ontario at that time. Since spruce is quite capable of tolerating perma-frost conditions, it is conceivable that the pioneer vegetation of southwestern Ontario may have resembled that now growing in the forest-tundra zone in the District of Keewatin, Northwest Territories, as described by Larsen (1965).

DISCUSSION

When the postulated Late-Wisconsin history of vegetation in southwestern Ontario is considered in conjunction with the current revised progress of deglaciation, the following sequence of events can be proposed as a possible working hypothesis.

All of southern Ontario was covered by ice during the Nissouri Stadial - the main and most extensive advance of Late-Wisconsin ice in this region (Dreimanis, 1977).

In the Erie Interstadial (about 16,000 - 15,500 years BP) much of southwestern Ontario was probably deglaciated (Mörner and Dreimanis, 1973) and pro-glacial lakes occupied the Lake Erie basin, including a postulated lower level phase (Lake Leverett). Although some migration of vegetation could have occurred into the central part of southwestern Ontario that was not covered by glacial lakes or ice, there is no evidence to indicate that spruce was present in this vegetation.

The ice readvance during the Port Bruce Stadial (about 15,500 - 13,500 years BP) occupied all of Lake Erie basin and extended south and north from it to deposit the Port Stanley Till. It appears that southwestern Ontario was again covered by ice, with the possible exception of small areas where the Catfish Creek Till of Nissouri Stadial age is exposed at the surface (Cowan et al., 1975). However, the ice began to retreat soon and according to studies by Barnett (1978, 1979) this retreat proceeded rapidly and resulted in a low-water phase (Lake Ypsilanti) in the Lake Erie basin, prior to another re-advance that deposited the Wentworth Till (Fig. 2) and the Paris and Galt Moraines.

It is possible that a major proportion of migration of the pioneer vegetation into southwestern Ontario occurred in late Port Bruce Stadial, especially during low-water phases in Lake Erie basin (Fig. 4), and during the following Mackinaw Interstadial (about 13,500 - 13,000 years BP). Initial populations of spruce could have been established in central southwestern Ontario at that time.

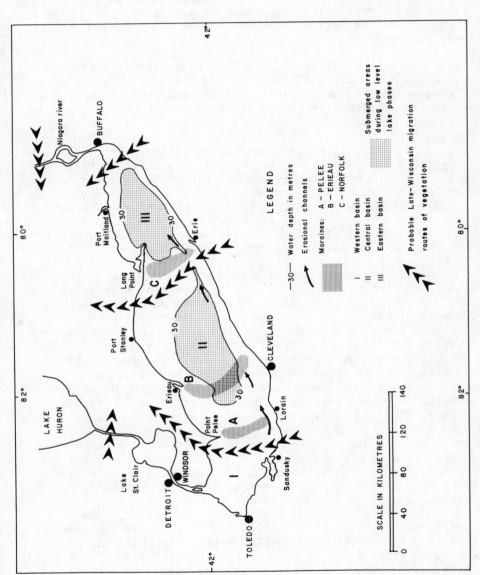

Fig. 4. Late-Wisconsin migration routes of vegetation across the Lake Erie basin during low-water stands, and the position of Port Bruce Stadial end moraines.

The readvance of ice during the Port Huron Stadial (about 13,000 years BP) deposited the Halton Till and coincided with the glacial Lake Whittlesey phase in the Lake Erie basin. There is no compelling reason to believe that tundra vegetation generally and some founder populations of spruce in ecologically favourable areas could not have survived the Port Huron ice readvance in southwestern Ontario.

A major ice retreat during the Two Creeks Interstadial resulted in the low level Early Lake Erie phase in the Lake Erie basin (Lewis, 1969) about 12,600 years BP, and this allowed again major migration of vegetation across the Lake Erie basin (Fig. 4) into southwestern Ontario.

If the outline of the above reasoning (that admittedly still requires confirmation) is essentially correct, it would help to explain the rather high percentages of spruce pollen found in the Late-Wisconsin herb pollen zone in southwestern Ontario prior to 12,500 radiocarbon years ago. It would also make it easier to account for the early presence of spruce in the Toronto area (based on macrofossil studies) if spruce was present in central southwestern Ontario during the Port Huron Stadial.

Another interesting implication of this study might be related to the Late-Wisconsin occurrence of mastodon in southwestern Ontario (Dreimanis, 1967, 1968). According to Dreimanis (1967) most of the fossil mastodon sites (a total of 50 - 60) are south of the glacial Lake Warren shoreline, and a few are in the central part of southwestern Ontario that was not covered by ice during the Port Huron Stadial. Dreimanis alludes to the possibility that mastodons could have lived in Ontario during the Erie Interstadial, and he states that "spruce is associated with mastodons as the principal indicator of their habitat" (Dreimanis, 1967, p. 669). There appears to be a commonly accepted association of mastodon with open spruce muskeg or woodland, and palynological studies in southwestern Ontario seem to support this assumption. Dreimanis (1967) postulated that a combined climate - vegetation change caused the extinction of mastodon in southern Ontario when the open spruce forest was replaced by a pine-hardwood forest.

Perhaps one of the rather obvious questions concerns the failure of mastodons to move northward with the migrating spruce forest to northern Ontario. The present study might offer a possible explanation. If, in fact, the mastodon lived in an open spruce forest environment, and was ecologically restricted to it, then conditions would have been quite suitable for mastodon in southwestern Ontario until the end of Port Huron Stadial.

However, at that time the major retreat of ice presumably coincided with some improvement of climate that allowed the development of closed spruce forest in the central part of southwestern Ontario that may have become unsuitable as a habitat for mastodon. The movement of mastodon to the east, west, and north would have been prevented for some time by the barrier of glacial lakes that effectively surrounded central southwestern Ontario, and the only alternative may have been a southward migration because at about that time the water levels in Lake Erie basin were rapidly dropping to the low Early Lake Erie stand and the large areas of exposed glacial lake sediments apparently supported an open spruce muskeg type of vegetation that probably was a satisfactory habitat for the mastodon.

During the Early Lake Erie phase, mastodon could have migrated (and probably did) into southwestern Ontario also from the region south of Lake Erie, and the northern limit of this migration would coincide with the glacial Lake Warren (and Lake Whittlesey) shorelines because this would have been the approximate southern boundary of the closed spruce forest mixed with jack pine.

Furthermore, studies by Roosa (1977) indicate that Paleoindian occupations probably coincided with the presence of mastodon as well as caribou in southwestern Ontario. It is likely that the Paleoindian provided significant assistance in the demise of the mastodon, but both may have found themselves in an environmental squeeze when the spruce forests closed and subsequently changed to a pine-hardwood forest. At that point in time there would not have been any ready environmental 'escape routes' from southwestern Ontario available for either the mastodon or the Paleoindian and both could have disappeared from southwestern Ontario at about the same time and for related reasons that at least in part were geobotanical.

The writer is not unaware of recent studies, reviews, and revisions of Late-Wisconsin deglaciation, environmental changes and history of vegetation south of the Lake Erie basin (for example, Calkin and Miller, 1977; Crowl, 1980; Miller, 1973; Muller, 1977) but a discussion of probable correlations has been omitted on purpose because it would require considerably more space to cover these matters in sufficient detail than intended within the limited scope of this report.

REFERENCES

Anderson, T. W. 1971. Postglacial vegetative changes in the
    Lake Huron - Lake Simcoe district, Ontario, with special
    reference to Glacial Lake Algonquin. Unpublished Ph.D.
    Thesis, University of Waterloo, Ontario, 229 p.

Barnett, P. J. 1978. Quaternary geology of the Simcoe area,
    southern Ontario. Ontario Division of Mines, Geological
    Report n. 162, 74 p.

--------. 1979. Glacial Lake Whittlesey: the probable ice
    frontal position in the eastern end of the Lake Erie bason.
    Canadian Journal of Earth Sciences, volume 16, pp. 568-574.

Calkin, P. E. and Miller, K. E. 1977. Late Quaternary environ-
    ment and man in western New York. Annals of the New York
    Academy of Sciences, volume 288, pp. 297-315.

Chapman, L. J. and Putnam, D. F. 1966. The physiography of
    southern Ontario (second edition). University of Toronto
    Press, Toronto, Ontario, 386 p.

Cowan, W. R., Karrow, P. F., Cooper, A. J. and Morgan, A. V.
    1975. Late Quaternary stratigraphy of the Waterloo - Lake
    Huron area, southwestern Ontario. Field Trips Guidebook
    (Waterloo '75), Part B, pp. 180-222, University of
    Waterloo, Waterloo, Ontario.

Cowan, W. R., Sharpe, D. R., Feenstra, B. H. and Gwyn, Q. H. J.
    1978. Glacial geology of the Toronto - Owen Sound area.
    Geological Association of Canada, (Toronto '78) Field
    Trips Guidebook (edited by A. L. Currie and W. O.
    Mackasey), pp. 1-16.

Crowl, G. H. 1980. Woodfordian age of the Wisconsin glacial
    border in northeastern Pennsylvania. Geology, volume 8,
    pp. 51-55.

Dreimanis, A. 1967. Mastodons, their geologic age and extinc-
    tion in Ontario, Canada. Canadian Journal of Earth
    Sciences, volume 4, pp. 663-675.

--------. 1968. Extinction of mastodons in eastern North
    America: testing a new climatic - environmental hypothesis.
    Ohio Journal of Science, volume 68, pp. 257-272.

Dreimanis, A. 1969. Late-Pleistocene lakes in the Ontario
    and the Erie basins. Proceedings of the 12th Conference,
    Great Lakes Research, pp. 170-180.

------- and Karrow, P. F. 1972. Glacial history of the Great
    Lakes - St. Lawrence region: the classification of the
    Wisconsinan stage, and its correlations. 24th International
    Geological Congreee (Montreal), Proceedings, Section 12,
    pp. 5-15.

------- and Goldthwait, R. P. 1973. Wisconsin glaciation in
    the Huron, Erie, and Ontario lobes. Geological Society of
    America, Memoir 136, pp. 71-106.

--------. 1977. Late Wisconsin glacial retreat in the Great
    Lakes region, North America. Annals of the New York
    Academy of Sciences, volume 288, pp. 70-89.

Hough, J. L. 1958. Geology of the Great Lakes. University of
    Illinois Press, Urbana, Illinois, 313 p.

--------. 1963. The prehistoric Great Lakes of North America.
    American Scientist, volume 51, pp. 84-109.

Karrow, P. F. 1963. Pleistocene geology of the Hamilton - Galt
    area. Ontario Department of Mines, Geological Report
    no. 16, 68 p.

-------, Anderson, T. W., Clarke, A. H., Delorme, L. D. and
    Sreenivasa, M. R. 1975. Stratigraphy, paleontology, and
    age of Lake Algonquin sediments in southwestern Ontario,
    Canada. Quaternary Research, volume 5, pp. 49-87.

Larsen, J. A. 1965. The vegetation of the Ennadai Lake area,
    N.W.T.: Studies in subarctic and arctic bioclimatology.
    Ecological Monographs, volume 35, pp. 37-59.

Lewis, C. F. M., Anderson, T. W. and Berti, A. A. 1966.
    Geological and palynological studies of Early Lake Erie
    deposits. Great Lakes Research Division, University of
    Michigan, Publication no. 15, pp. 176-191.

Lewis, C. F. M. 1969. Late-Quaternary history of lake levels
    in the Huron and Erie basins. Proceedings of the 12th
    Conference, Great Lakes Research, pp. 250-270.

McAndrews, J. H., Boyko, M. and Byrne, R. 1974. Investigations
    at Crawford Lake. Geological and vegetational history.
    Friends of the Pleistocene 37th Annual Reunion (Toronto),
    May 1974, 10 p.

Miller, N. G. 1973. Late glacial plants and plant communities
    in northwestern New York State. Journal of the Arnold
    Arboretum, volume 54, pp. 123-159.

Morgan, A. V. 1972. Late Wisconsin ice-wedge polygons near
    Kitchener, Ontario, Canada. Canadian Journal of Earth
    Sciences, volume 9, pp. 607-617.

Mott, R. J. and Farley-Gill, L. D. 1978. A Late-Quaternary
    pollen profile from Woodstock, Ontario. Canadian Journal
    of Earth Sciences, volume 15, pp. 1101-1111.

Muller, E. H. 1977. Late glacial and early postglacial
    environments in western New York. Annals of the New York
    Academy of Sciences, volume 288, pp. 223-233.

Mörner, N.-A. and Dreimanis, A. 1973. The Erie Interstade.
    Geological Society of America, Memoir 136, pp. 107-134.

Prest, V. K. 1970. Quaternary geology of Canada. In: Geology
    and economic minerals of Canada (R. J. E. Douglas, editor);
    Geological Survey of Canada, Economic Geology Report no. 1,
    5th edition, Chapter 12, pp. 676-764.

Roosa, W. B. 1977. Great Lakes Paleoindian: the Parkhill
    site, Ontario. Annals of the New York Academy of Sciences,
    volume 288, pp. 349-354.

Sly, P. G. and Lewis, C. F. M. 1972. The Great Lakes of
    Canada - Quaternary geology and limnology. 24th Inter-
    national Geological Congress (Montreal), Field Excursion
    A43 Guidebook, 92 p.

Sreenivasa, B. A. 1973. Paleoecological studies of Sunfish
    Lake and its environs. Unpublished Ph.D. Thesis, Uni-
    versity of Waterloo, Ontario, 184 p.

Terasmae, J., Karrow, P. F. and Dreimanis, A. 1972. Quaternary
    stratigraphy and geomorphology of the eastern Great Lakes
    region of southern Ontario. 24th International Geological
    Congress (Montreal), Field Excursion A42 Guidebook, 75 p.

------- and Matthews, H. L. (1980). Late-Wisconsin white spruce
    (Picea glauca (Moench) Voss) at Brampton, Ontario. Canadian
    Journal of Earth Sciences, Vol. 17, pp. 1087-1095.

Winn, C. E. 1977. Vegetational history and geochronology of
    several sites in south southwestern Ontario with discussion
    on mastodon extinction in southern Ontario. Unpublished
    M.Sc. Thesis, Brock University, Ontario, 373 p.

THE PALEOLIMNOLOGY OF ROSE LAKE, POTTER CO., PENNSYLVANIA: A COMPARISON OF PALYNOLOGIC AND PALEO-PIGMENT STUDIES

James F. P. Cotter, Dept. of Geological Sci., Lehigh
Univ., Bethlehem, Pennsylvania 18015, and G. H. Crowl,
Dept. of Geology & Geography, Ohio Wesleyan Univ.,
Delaware, Ohio 43015

ABSTRACT

Rose Lake is situated in a kettle in the Woodfordian border in Potter Co., north-central Pennsylvania. A 13-meter core collected in February, 1977 is the basis for stratigraphic palynology and paleolimnology reported here.

The pollen stratigraphy of Rose Lake sediments is comparable to that of the Middle Atlantic region. Palynologic evidence indicates a climate-induced vegetation succession from a spruce parkland (the A Pollen Zone of the New England Region) to a pine forest (B Zone) to a mixed hardwood forest (C Zone) in the sur-rounding region. Organic-rich sedimentation began 14,175 ± 100 years B.P. (SI 3098). These basal sediments are dominated by Picea (spruce) and Pinus (pine) Spp. pollen characteristic of the A_2 Zone. A distinct decline of hemlock (the close of the C_1 Sub-Zone) is believed to have been caused by biologic rather than climatic factors, indicating that little climatic change has occurred in the region since the deposition of the Pine-Oak Assemblage Zone (B Zone).

Temporal changes in the sedimentary environment, along with changes in flora of the lake have brought about differential sedimentation and preservation of paleo-pigments. Studies of the changes in total relative amounts of sediment, organic content, chlorophyll derivatives, and carotenoids indicate that Rose Lake was oligotrophic from its inception and remained

oligotrophic throughout its history. Productivity of the lake
gradually increased, decreased, and then increased again. These
changes do not correlate with changes in the pollen record,
therefore they were probably induced by changes in the relative
availability of nutrients in the lacustrine system, or by filling
of the lake basin with sediment. Superimposed on the general
trends of productivity increase, decrease, increase, are two
brief episodes of oxidation of profundal sediments indicating
either rapid overturn of the lake waters or oxidizing conditions
throughout the lake. These conditions may have been caused by
changes in regional runoff patterns.

Comparison of the pollen and pigment stratigraphies of Rose
Lake indicates that climate is not the determining factor in
changes in lake productivity. However, paleopigment studies
in conjunction with palynologic studies provide information pertaining to the reconstruction of the history of the lacustrine
system and surrounding region.

INTRODUCTION

Rose Lake is a small kettle hole lake situated near the
Woodfordian ice limit in Potter County, Pennsylvania (Figure 1).
A 13-meter core collected in February, 1977 is the basis for
stratigraphic palynology and paleolimnology. Through a comparative reconnaisance study of the Late- and Post-Glacial fossil
pigment and pollen records of the Rose Lake core, the history
of the climatic and biologic factors which have influenced the
lake and the surrounding region has been reconstructed. In
addition, short-lived events, or events with limited influence
have also been recognized in the paleopigment and pollen records
which might not have been evident if a comparative study were
not implemented.

Sedimentary fossil pigments, chlorophyll derivatives, and
carotenoids have been studied for years as biochemical fossils.
In the reconstruction of the paleoecology of lacustrine systems,
fossil pigments are especially valuable where microfossils are
not always sensitive or representative indicators. They are
particularly useful as indices of past and present trophic
status (Vallentyne, 1956, 1957, 1969; Brown, 1969; Gorham, 1960,
1961; Gorham and Sanger, 1972; Sanger and Gorham, 1970, 1972,
1973; and Sanger and Crowl, 1979). Sedimentary pigments can
also be used to determine changes in the balance between autochthonous and allochthonous organic sediment contribution
(Gorham and Sanger, 1975), the source of sedimentary organic
matter (Zullig, 1961; Brown and Colman, 1963; Griffiths, et al.,
1969; Griffiths and Edmundson, 1975; and Gorham and Sanger, 1976),

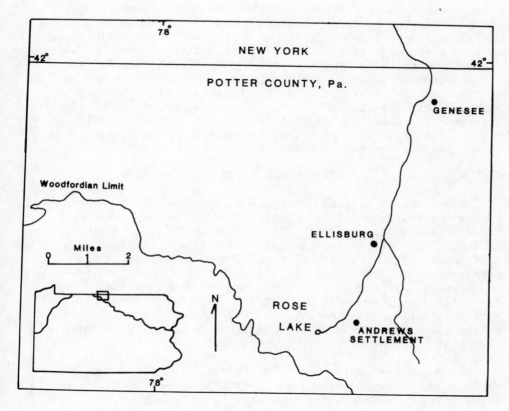

Fig. 1.   Location map of the Rose Lake site.

and the presence or absence of open water conditions throughout
the history of individual lakes (Fogg and Belcher, 1961; Belcher
and Fogg, 1964; Czeczuga, 1965; Sanger and Gorham, 1972; Sanger
and Crowl, 1979).

Since the study by Deevey (1939), numerous studies of the
Late-Glacial pollen record in the northeastern United States
have provided a well-documented record of climatically and
biologically induced forest succession. The consistent similarities in this record make possible the correlation of the Late-
and Post-Glacial pollen assemblage zones of the northeast. Comparison of the pollen and sedimentary pigment stratigraphies of
Rose Lake enables the distinction of climatically induced changes
in the lacustrine system (recorded by the pollen record) from
changes induced by other factors (e.g., trophic status and
nutrient availability) recorded by the fossil pigment record.
Although climatically induced changes in the pollen stratigraphy
did not correspond to changes in the pigment record, studies
showed that regional vegetation changes did influence the fossil
chlorophyll derivatives and carotenoid record.

SITE DESCRIPTION

Rose Lake is located 1.5 km west of Andrews Settlement,
Potter County, Pennsylvania, and 11.4 km south of the New York
border (Figure 1), on the glaciated portion of the Appalachian
Plateau. The Woodfordian glacial border (Lewis, 1884; Denny,
1956; Crowl, 1980; Crowl and Sevon, 1980), deposited by the
Ontario-Hudson Lobe of the Laurentide Ice Sheet, extends from
the Delaware River near Belvidere, New Jersey, to the apex of
the Salamanca re-entrant, 5 km north of the New York-Pennsylvania
border in Cattaraugus County, New York. Rose Lake was formed in
an ice block depression on the "Terminal Moraine", 2 km north of
the Woodfordian ice limit in a pre-glacial valley.

Rose Lake is approximately 5.2 hectares in area and 6.6 m
at its deepest point. The lake is situated 690 m above sea
level near the divide which separates the drainage of the
Genesee River and the Allegheny River. Rose Lake is oligotrophic
with a single outlet, Rose Lake Run, a tributary of the West
Branch of the Genesee River.

Three distinct forest regions are represented in Potter
County, Pennsylvania (Goodlett, 1954): an Oak Forest region, a
Northern Hardwood Forest region and a Transition region.
Küchler (1964) considers this region part of the "Conifer-
Hardwood Forest" of North America. At present, the forests
growing in the drainage of the Genesee and Allegheny Rivers

consist of approximately 25 species of trees. The area surrounding Rose Lake is predominantly cleared farmland with
scattered patches of mixed hardwood forest dominated by Tsuga
canadensis (eastern hemlock).

METHODS

Thirteen meters of sediment core were obtained in one meter
increments with a modified Livingston corer in February, 1977,
when the lake was frozen. Core sections were stoppered in the
field and stored in a cold room until extrusion and sectioning
for laboratory studies. Laboratory analysis consisted of a
reconnaissaince survey of fossil pigments and pollen for this
study. Samples for fossil pigment analysis were taken at .5 m
intervals and the sampling interval for pollen studies was .3 m,
with additional samples taken near pollen zone boundaries.

Chemical preparation of pollen samples included defloccula-
tion with KOH, leaching of silicates with HF, and staining with
gentian violet as modified from the methods of Erdtman (1943).
Slides were made from stained pollen concentrates suspended in
glycerin gelatin.

Identification of pollen grains was aided by reference keys
of Erdtman (1952), Faegri and Iverson (1964), McAndrews, et al.
(1973), Moore and Webb (1978), and Basset et al. (1978). Pollen
counts were made at a magnification of 450x and detailed examination of individual grains at 1000x with an oil immersion lens
on a Zeiss binocular microscope.

Pollen percentages are based on counts of 400 grains of
both arboreal pollen (AP) and non-arboreal pollen (NAP). Unknown
pollen grains, and grains unrecognizable due to corrosion,
crumpling, or obscurring detritus were counted and grouped as
"unknown" pollen, but were not included in the pollen sum.
Sphagnum, Lycopodiaceae, Polypodiaceae and other spores were
counted separately, and were also not included in the pollen
sum.

Pinus spp. (pine) were separated into two groups: Haploxylon
(Pinus strobus-type) and "others"; based on the morphologic
methods of Ueno (1958) and McAndrews et al. (1973). However,
Klaus (1978) has stated that the verrucae of the cappula-nexine,
although always developed on species of the sub-genus strobus,
are also encountered on species not of that sub-genus. Hence,
differentiation of these groups must be considered tentative.

The grouping of pollen types was necessary where problems in genera distinction exist. The betuloid group includes Betula (birch), Carpinus (hornbeam), Ostrya (hop-hornbeam), and Corylus (hazelnut). The inaperaturate gymnosperm pollen of Thuja (arborvitae), Juniperus (juniper) and Chameacypris (southern white cedar) are included in the TJC group because juniper, which has a rudimentary pore, could not be distinguished from the other two genera.

Pigments were extracted, and percent organic matter was determined using the methods described by Sanger and Gorham (1972). Pigments were measured spectrophotometrically using a slit band of less than 0.06 nm for carotenoids at the 445 to 450 nm peak and chlorophyll derivatives at the 664 to 666 nm peak. Pigment concentrations are expressed as spectrophotometric units per gram organic matter; one unit is equivalent to an absorbance of 1.00 in a 10 cm cell when dissolved in 100 ml of solvent.

For pigment diversity determination, thin-layer chromato-graphic plates were coated with silica-gel 7G (Baker Chemical Co.), spotted with a pigment solution, and developed. Pigment diversity is expressed as total spot number on thin-layer chromatograms.

CORE DESCRIPTION

Basal sediments of the Rose Lake core (13.0 to 11.8 m) are predominantly clays (Figure 2). These organic-deficient clays are characteristic of very early Post-Glacial lacustrine sedi-mentation in this region, and represent the period between lake formation and the establishment of vegetation in the lake and the surrounding area. Above these basal clays (11.8 to 11.0 m, Figure 2) is a layer of green gyttja which contains irregularly spaced clay lenses. The deposition of the gyttja marks the attainment of relatively high rates of production and deposition of organics in the lacustrine system.

Overlying the green gyttja is a distinct zone of laminated clay and gyttja (11.0 to 9.0 m), with individual laminae that vary in thickness throughout the horizon. The laminae probably represent episodic changes in the character of lacustrine pro-ductivity, or sedimentation, such as increased erosion during periods of heavy rain. The cause and periodicity of these changes however, has not been determined.

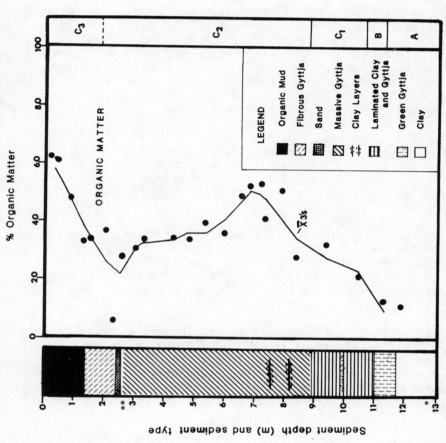

Fig. 2. Percent organic matter profile in the Rose Lake core. The curve is a running average of threes. Radiocarbon samples are indicated: **2,525 ± 110 and *14,175 ± 100 years B.P.

The massive gyttja overlying the laminated clay and gyttja
(9.0 to 5.2 m) represents a marked change in the character and
origin of the sediments of Rose Lake. Gyttja generally is deposited when algae and other aquatics are the major contributors
of pond sediments rather than peatland species (Sanger and Crowl
1979). During the deposition of this unit, productivity in the
lake reached a maximum (discussed later) and organic sedimentation was relatively rapid.

The deposition of the massive gyttja was interrupted twice
by deposition of organic-rich mud layers at 8.2 and 7.6 m and
deposition of layers of wood and leaf fragments at 7.2 and 5.8 m.
The organic-rich muds probably represent brief episodes of low
algal productivity resulting in less organic sedimentation and
the leaf and wood fragments probably represent brief periods of
excessive erosion of an incipient peatland, or high input of wind-
or runoff-transported terrestrial organic debris.

A layer of fine sand at 2.5 m represents an interruption of
organic deposition by rapid clastic deposition. This sand
horizon (2.5 m to 2.3 m) is massive with no sedimentary
structures. A preliminary examination showed numerous seeds of
Potamogeton and other pond weeds, and wood fragments within this
horizon.

Following the deposition of the fine sand layer, brown,
fibrous gyttja was deposited (2.3 to 1.3 m) indicating a continuously developed and eroded peatland was established on the
margins of Rose Lake. Overlying the brown gyttja, to the surface
of the core, are poorly compacted dark lake muds with some clay
horizons.

Organic matter in the Rose Lake core (Figure 2) increases
from very low concentrations in the basal clay horizon to 45% in
the massive gyttja (6.8 m). From 6.8 m where this maximum is
reached, percent organic matter begins to decrease until 2.5 m,
the fine sand horizon, where organic matter is at its lowest
percentage. From 2.5 m to the top of the core, percent organics
increase again at a relatively rapid rate.

As vegetation became established in Rose Lake and the surrounding area, percent sedimentary organic matter gradually increased. As the available nutrient supply was utilized and then
isolated through sediment burial, nutrient supply probably became a limiting factor to biotic productivity. The decrease in
percent organic matter from 6.7 m to 2.5 m is believed to be
the result of the decrease of available nutrients. After the

deposition of the fine sand layer, it appears a new source of
nutrients was again made available to the lake system. The
source of this nutrient supply is difficult to discern.

The cause of the deposition of the sand layer and the subsequent increase in the available nutrient supply is problematic.
Possibly local forest fires not reflected in the pollen record
(discussed below) resulted in increased runoff and the deposition
of coarser sediments; but no charcoal was found in this horizon.
Abundant Potamogeton seeds found in the sand layer suggest that
this may be a slumped or reworked near-shore facies, but no
deformation structures were recognized and slumping would not
explain the subsequent increase in lacustrine productivity. The
deposition of the sand layer then, remains enigmatic until both
its cause and its effect on the lake system can be explained.

Two radiocarbon dates have been obtained from the Rose Lake
core. The oldest, 14,175 ± 100 yrs. B.P. (SI-3098) provides a
minimum date for the deglaciation of this area. This date obtained from the base of the organic-rich section of the core is
correlative with the regional A_2 pollen zone of Davis (1969) and
is one of the oldest minimum dates for the retreat of ice from
the maximum Woodfordian position in Pennsylvania (Crowl, 1980).

The second date, 2,525 ± 110 (SI-4328), was obtained from
just below the sand horizon (2.4 m). This date provides a
minimum date for the re-establishment of Tsuga (hemlock) in the
region (discussed later).

RESULTS AND DISCUSSION

Palynology

The Late- and Post-Glacial pollen assemblage stratigraphy of
Rose Lake (Figure 3) is similar to pollen stratigraphies in New
England (Davis, 1965, 1969, 1976; Whitehead, 1979), New York
(Sirkin, 1967, 1976, 1977; Miller, 1973), and Pennsylvania
(Martin, 1958; Sirkin, 1977). The pollen assemblage zones
designated for the Rose Lake core are correlated to the classic
T, A, B, and C Regional Pollen Zones of Deevey (1949) and
Davis (1965) primarily for ease of description. Interpretation
of paleo-vegetation and climatic characteristics from pollen
assemblage zones is based primarily on comparison with isopollen
lines (Davis and Webb, 1975) and pollen diagrams of other workers
(e.g., Sirkin, 1977; Whitehead, 1979).

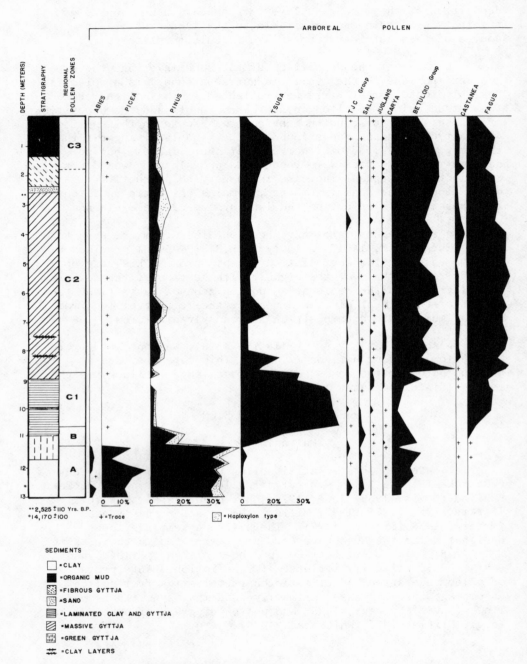

Fig. 3. Pollen diagram of Rose Lake. For a full-size version of
         this illustration, see fold-out.

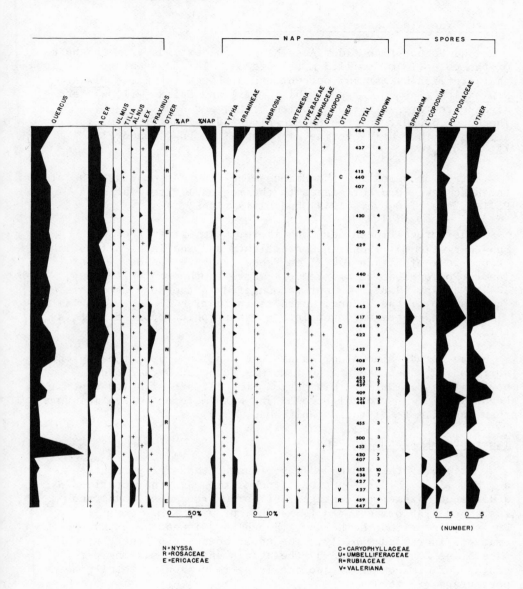

N=NYSSA
R=ROSACEAE
E=ERICACEAE

C=CARYOPHYLLACEAE
U=UMBELLIFERACEAE
R=RUBIACEAE
V=VALERIANA

Spruce-Pine-Alder Assemblage Zone

   The Spruce-Pine-Alder Assemblage Zone extends from the base
of the core (13 m) to 11.3 m (Figure 3). This assemblage zone is
characterized by maximum percentages of Picea (spruce), Pinus,
Diploxylon (red-pine type), Abies (fir) and Salix (willow), and
high percentages of Betula (birch group) and Alnus (alder).
The upper boundary of the Spruce-Pine-Alder Zone (11.3 m) is
defined by the rapid decline of spruce (19% to 0%) and fir
(3% to 0%).

   Stratigraphically, the Spruce-Pine-Alder Assemblage Zone of
the Rose Lake core correlates to the regional Late-Glacial A
Pollen Zone of Deevey (1949) and Davis (1965).

   The Spruce-Pine-Alder Zone is very similar to the A_2, A_3
Sub-Zone of Davis (1969) in Connecticut, and the later half of the
Spruce Zone of Sirkin (1977) in Pennsylvania. The Herb, or
Tundra Pollen Zone recognized elsewhere in the region (Deevey,
1949; Martin, 1958), was not encountered in the Rose Lake core.
The absence of the T Zone, which is indicative of a periglacial
climate, indicates an incomplete stratigraphic record from this
core.

   The Spruce-Pine-Alder Assemblage Zone indicates the region
surrounding Rose Lake at the time was occupied by a closed
boreal forest. During this period, the climate of north-central
Pennsylvania was probably cooler and with more effective moisture
than the climate of today (Whitehead, 1979).

The Pine-Oak Assemblage Zone

   Following the rapid decline of spruce and fir percentages,
a short (11.3 - 10.8) period of deposition was characterized by
a distinct maximum percentage (35%) of Quercus spp. (oak) and
sustained high percentages (15 - 20%) of pine (Diploxylon).
During this period, both Fagus (beech) and Tsuga (hemlock) percentages increased rapidly. The upper boundary of the Pine-Oak
Assemblage Zone is marked by the attainment of a maximum percentage of hemlock (65%) and the large reduction of total pine
percentage (23 to 4%).

   The Pine-Oak Zone is here correlated to the regional B-Zone
of Deevey (1949) and Davis (1965). Like the Spruce-Pine-Alder
Zone, the Pine-Oak Zone of the Rose Lake core is similar to
assemblages recognized at individual locations elsewhere in
northeastern North America (e.g., Davis, 1969). The Pine-Oak
Assemblage Zone (as does the B Zone elsewhere) probably represents earliest Post-Glacial deposition in this area of
Pennsylvania.

Because oak spp. may be over-represented in the Pine-Oak
Assemblage Zone as an artifact of the "percentage-type" diagram,
the vegetation of the Rose Lake region was probably a mixed
conifer and deciduous forest during this period. This type of
forest is analogous to the present day vegetation northeast of
Lake Huron (Whitehead, 1979). Climatic conditions were drier
and probably cooler during the deposition of the Pine-Oak
Assemblage Zone than present conditions in northern Pennsylvania.

The Oak-Hemlock-Beech-Birch Assemblage Zone

The remainder of the Rose Lake core, from 10.6 m to the
surface, is characterized by varying percentages of hemlock and
the predominant mixed hardwood species, such as oak, beech and
birch, while coniferous species continued to decrease in representation in the pollen record. This assemblage zone is correlated
to the C-Zone in New England (Deevey, 1949; Davis, 1965) and the
Mid-Atlantic region (Sirkin, 1977). The Oak-Hemlock-Beech-Birch
Assemblage Zone of the Rose Lake core has been divided into three
assemblage sub-zones for ease of discussion, the Hemlock, the
Maple-Beech-Birch, and the Hemlock-Chestnut-NAP Assemblage Sub-
Zones. These sub-zones have been correlated to the C_1, C_2, and
C_3 regional sub-zones of Deevey (1949) and Davis (1965).

Hemlock Assemblage Sub-Zone

The base of the Hemlock Assemblage Sub-Zone is defined by a
hemlock percentage maximum following the appearance and rapid
increase of hemlock. Along with the high percentages of hemlock
throughout the sub-zone, this period is characterized by low percentages of pine and increasing proportions of birch and beech
pollen. The establishment of the hemlock-dominated hardwood
forest represented by this assemblage sub-zone indicates that
climatic conditions in this part of Pennsylvania were similar to
those of today.

The upper boundary of the Hemlock Assemblage Sub-Zone (8.7 m)
is marked by a rapid decrease in the percentage of hemlock (from
39 to 12%). A similar rapid decline in hemlock is recognized
throughout northeastern North America (Davis, this volume) in
both "absolute" and percentage type pollen diagrams. Because of
the relative short period in which hemlock declined over such a
large region, Davis (this volume) has suggested that biologic
rather than climatic factors induced vegetation change. Likens
and Davis (1975) and Davis (this volume) have suggested that a
disease or insect may have preferentially attacked hemlock, and
because of a lack of natural resistance, this major forest species
was nearly eradicated from North America.

This decline in hemlock throughout the northeast occurred at approximately 4,800 ^{14}C yrs. B.P. with little latitudinal variation (Davis, this volume). Until resistant strains of hemlock developed, hemlock growth in North America was suppressed. In the Rose Lake region, initial stages of hemlock re-establishment began after 2,525 ± 110 yrs. B.P. (2.6 m, Figure 3). Assuming the initial decline occurred at 4,800 yrs. B.P., the suppression of hemlock growth lasted approximately 2,300 ^{14}C yrs. in this portion of Pennsylvania.

Maple-Beech-Birch Assemblage Sub-Zone

Following the decline of hemlock in the Rose Lake region, percentages of Castanea (chestnut), Carya (hickory), Acer (maple), birch and beech increased rapidly. This period has been designated the Maple-Beech-Birch Assemblage Sub-Zone. The Maple-Beech-Birch Assemblage Sub-Zone is a correlative of the C$_2$ Sub-Zone in New England. Although previous studies have suggested the C$_2$ Sub-Zone may represent cooler and moister climatic conditions then at present, if the regional decrease in hemlock is due to biologic factors the relative increase in percentages of other hardwood species represents the natural successional process of hemlock replacement. Thus, in the Rose Lake core, little evidence for climatic change is recognized at the Hemlock/Maple-Beech-Birch Assemblage Sub-Zone boundary (8.6 m). A climate similar to the present day climate probably existed through the transition, and in turn, throughout the deposition of the remainder of the Oak-Hemlock-Beech-Birch Assemblage Zone.

Hemlock-Chestnut-NAP Assemblage Sub-Zone

The lower boundary of the final sub-zone in the Oak-Hemlock-Beech-Birch Assemblage Zone is tentatively placed at 1.8 m. At this level; hemlock percentages again begin to increase and a maximum percentage of chestnut pollen is attained. This sub-zone represents the re-establishment of hemlock as the dominant forest member in the region probably due to the development of resistant strains of hemlock.

The uppermost portion of the Hemlock-Chestnut-NAP Assemblage Sub-Zone is dominated by high percentages of non-arboreal species Graminaceae (grass) and Compositae, artemesia-type, indicative of land clearance associated with European settlement of the region.

Paleo-pigments

A general interpretation of the significance of fossil pigment studies was discussed by Sanger and Gorham (1972). It has been shown that chlorophyll derivatives and carotenoids, preserved in the sedimentary profile due to the absence of light, oxygen and warmth, reflect changes in the in situ production of organics. In the Rose Lake core, although the sampling interval did not allow the recognition of most short-lived events, paleo-pigment studies indicate the lake maintained a productivity level characteristic of oligotrophic lakes with little change throughout its history.

Concentrations of chlorophyll derivatives and carotenoids (Figure 4) directly reflect lacustrine productivity. A greater concentration of pigment at any given level in the sedimentary profile can be indicative of a higher rate of productivity at that time.

Pigment concentration curves of Rose Lake indicate productivity increased slightly from the inception of organic sedimentation until a peak was reached at approximately 7.5 m. From 7.5 m to 2.4 m, productivity in the lake decreased slightly, and from 2.4 meters to the core top productivity again increased. The rates of change of the lake productivity were slight in all three intervals, with the most rapid change occurring during the final stages of deposition (2.4 m to surface).

Superimposed on the productivity trends (slight increase, slight decrease, more rapid increase) are two brief periods of low productivity or poor pigment preservation evidenced by low pigment concentrations at 8.5 m and 2.2 m. The significance of these events is difficult to discern. The lowermost pigment decline, 8.5 m, is not represented in the percent organics curve (Figure 2) and therefore probably does not represent a decline in lake productivity. It is likely that there was a relative increase in allocthonous contribution, or profundal sediments were exposed to oxidizing conditions prior to burial, resulting in the destruction of sedimentary pigment.

The rapid decline in sedimentary pigment concentration at 2.2 m coincides with both a distinct low in percent organics and the sand horizon previously discussed (Figure 2). This minimum represents either an influx of sediment, or the degree of preservation of sedimentary pigments in profundal sediments. It may also be related to the changes in the characteristics of lacustrine sedimentation which resulted in the deposition of the sand.

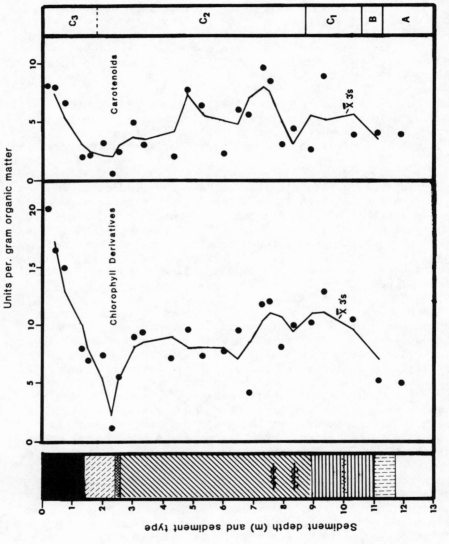

Fig. 4. Profile of sedimentary chlorophyll and carotenoids in the Rose Lake core.
The curves are running averages of threes. Sediment types are as in Fig. 2.

The ratio of chlorophyll derivatives to carotenoids indicates, to some degree, trophic status in the lacustrine environment; but more importantly, provides a good indication of relative importance of autochthonous versus allochthonous organic matter (Sanger and Crowl, 1979). Low values of the chlorophyll derivative to carotenoid ratio indicate greater autochthonous production and deposition of organics. This is due to the relatively high susceptibility of allochthonous carotenoids to oxidation prior to burial.

The values of the chlorophyll derivative to carotenoid ratio throughout the Rose Lake core indicate a high percentage of allochthonous contribution of sedimentary organics (Figure 5), and is characteristic of oligotrophic lakes where O_2 values remain high in bottom waters throughout the entire year. From the core base (13 m) to 6.8 m, ratios range from 1.28 to 3.82 with no clear indication of a general trend. At 6.8 m, the chlorophyll derivative vs. carotenoid ratio of the Rose Lake core attains a minimum value of 0.7 (characteristic values in oligotrophic lakes range from 0.5 - 1.0 vs. 0.3 to 0.5 in eutrophic lakes; Sanger and Crowl, 1979), indicating a low point in the deposition of autochthonous organics was attained. This ratio minimum roughly corresponds with the peak maximum in lake productivity recorded by pigment concentration curves (Figure 4).

From 6.8 m to 2.2 m, the chlorophyll derivative to carotenoid ratio shows a general trend of increasing allochthonous contribution following the peak in autochthonous deposition. This trend is associated with decreasing levels of lake productivity indicated by pigment concentration curves (Figure 4) and is probably attributable to a decline in available nutrients in the lacustrine system previously discussed.

At 2.2 m, again corresponding with the deposition of the sand horizon, there is a slight shift in the trend of chlorophyll derivatives to carotenoid ratio indicative of greater autochthonous organic deposition. This shift to increased autochthonous input of organics is also evidenced by an increase in productivity above 2.2 m which resulted in higher pigment concentrations (Figure 4). The increase in relative proportion of autochthonous organic sediments associated with the increase in lacustrine productivity further indicates a new or recycled nutrient supply was made available.

Due to the variability in the ratio of chlorophyll derivative to carotenoid ratio of the Rose Lake core, it is difficult to discern the importance of deviations from the general trends discussed above, since the lake was oligotrophic throughout its history. The short-lived events recognized in the pigment

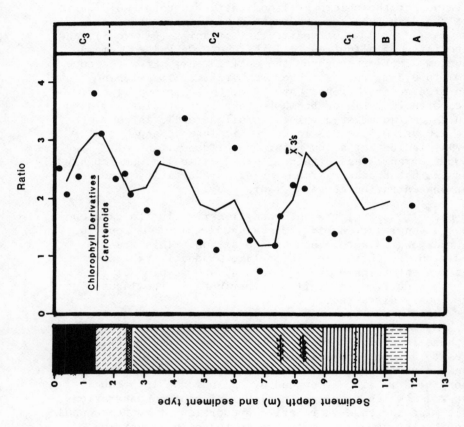

Fig. 5.   Profile of chlorophyll derivatives: carotenoids ratio in the Rose Lake core.
The curve is a running average of threes. Sediment types are as in Fig. 2.

concentration curves (Figure 4) are not clearly represented in the chlorophyll derivative to carotenoid ratio.

Pigment diversity is indicative not only of biotic diversity of the lacustrine system, but more importantly, redox conditions at the sediment/water interface. Because carotenoids are much more sensitive than chlorophyll and its derivatives to changes in water temperature, light influx, or redox conditions in profundal sediments, a fewer number of spots (lower pigment diversity) could indicate oxidizing conditions in the uppermost sediments. Oxidizing conditions could result from lower water stages, an increase in the periodicity of the overturn of the lake waters, or a low profundal BOD.

In the Rose Lake core, from the base to present-day sediments, spot numbers range from 10 to 15 and are indicative of oligo-trophic conditions throughout the history of the lake (Figure 6). The pigment diversity curve, however, shows evidence of two significant periods of extremely aerobic conditions at 8.2 to 9.0 m, and 2.4 to 1.5 m, both of which correlate with the short-lived event recognized in the pigment concentration curves (Figure 4). In addition, the upper peak (2.4 to 1.5 m) also correlates with the deposition of the sand layer previously discussed.

The two periods of low spot number in the Rose Lake core are believed to represent short-lived episodes of disturbance of profundal sediments. The earliest disturbance (9.0 m) had little effect on the lake system and may have been caused by increased runoff associated with the decline of hemlock in the region. This decline in a major forest member would have opened the forest significantly. The latter episode (2.4 m) was recorded both in the lacustrine system (pigment concentration and percent organic matter), and the sedimentary profile (the sand horizon). This disturbance also had long lasting effects on the lake, because as a result of this event, a new source of nutrients was provided to the lacustrine system, resulting in greater productivity (previously discussed).

CONCLUSIONS

The pollen record of Rose Lake is similar to the pollen records of much of northeastern North America. Three distinct pollen assemblage zones are recognized in the Rose Lake core, each representing a particular set of climatic conditions. Potter County, Pennsylvania had been deglaciated by 14,175 yrs. B.P. and at this time a closed boreal forest (Spruce-Pine-Alder Assemblage Zone) existed in the region, indicating the climate

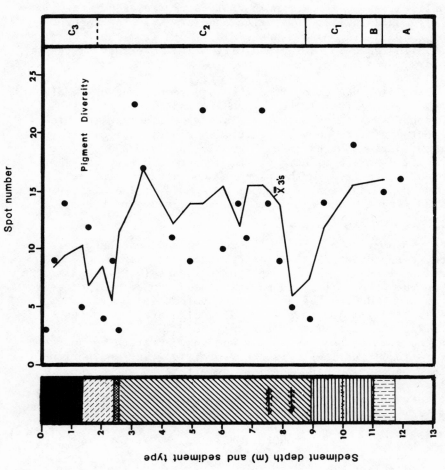

Fig. 6. Profile of the Pigment diversity ascertained from spot number on 2-dimensional thin-layer chromatograms from the Rose Lake core. The curve is a running average of threes. Sediment types are as in Fig. 2.

was colder and wetter than climatic conditions in this region today. The earliest Post-Glacial onset of dryer, but still cool conditions brought about the establishment of a mixed conifer and hardwood forest (Pine-Oak Assemblage Zone). The subsequent establishment of climatic conditions similar to those in the region today resulted in the establishment of a mixed hardwood forest (Oak-Hemlock-Beech-Birch Assemblage Zone) which existed in the region until Europeans settled in the area. During the period when mixed hardwood species prospered in the region, the expansion of hemlock was severely disrupted probably by biological factors that resulted in a distinct decline of hemlock, the opening of the forest, and the natural expansion and establishment of species that were previously of minor extent.

Fossil sedimentary pigment data indicate that Rose Lake changed very little throughout its history, remaining oligotrophic from its inception. Because climatic changes indicated by the pollen record do not coincide with changes in the paleo-pigment record, other factors such as nutrient availability and changes in the circulation patterns of the lake (possibly due to changes in tributary flow, rate of overturn, or filling of the basin), appear to be the dominant influence upon lake productivity.

Lacustrine productivity, as evidenced by percent organics and pigment concentrations, increased gradually. This gradual increase in lacustrine productivity continued through both the Spruce-Pine-Alder and the Pine-Oak Assemblage Sub-Zone. During the Beech-Birch-Maple Assemblage Sub-Zone, productivity in Rose Lake began to decline. It is believed that this decline was caused by a decrease in the lake's nutrient supply, the result of burial and isolation of previously used nutrients in organic sediments and the formation of stabilized soils and soil humus layers. The deposition of a sand layer at 2.4 to 2.2 m in Rose Lake was followed by an increase once again in productivity.

Superimposed upon this general trend of productivity increase, followed by decrease, followed by more rapid increase are two brief periods of oxidation of the pigments of profundal sediments or an increase in deposition of oxidized allocthonous material. These episodes may be due to increases in runoff resulting in rapid overturn of the lake, oxidizing conditions throughout the lake, or the deposition of large concentrations of allochthonous detritus.

The first of these two events occurred near the close of the Hemlock Assemblage Sub-Zone. During this time, hemlock, the principal component of the surrounding forest, was severely reduced. This reduction would have resulted in the opening of

much of the forest, an increase in forest detritus and an increase in regional runoff. It is the decline of hemlock, therefore, that is believed to have resulted in the first brief period
of sedimentary pigment oxidation.

   The second episode of oxidation corresponds with the deposition of the sand layer. Although the origin of this sand
layer is undetermined, its deposition is believed to be related
to an increase in local runoff, perhaps due to small forest fires
which did not affect the pollen record. The increase in productivity subsequent to the deposition of the sand layer implies
that a nutrient supply had once again become available through
erosion of the landscape.

   Comparison of the pollen and pigment stratigraphies of Rose
Lake indicates that climate may not be the determining factor in
changes in lake productivity in these moist temperate regions.
However, paleopigment studies in conjunction with palynologic
studies provide information pertaining to the reconstruction of
the history of lacustrine systems and surrounding regions.

ACKNOWLEDGEMENTS

   We thank Robert Stuckenrath of the Smithsonian Institution
for providing radiocarbon dates, and William Hill and Ruth Hay
for aid in preparation of pollen samples and pigment separation.
We are grateful to Jon Sanger and Les Sirkin for helpful comments
on the manuscript.

REFERENCES CITED

Basset, I. J., Crompton, C. W., and Parmalee, J. A. 1978. An
    atlas of airborn pollen grains and common fungus spores of
    Canada. Can. Dept. Agriculture Monograph No. 18, 321 p.

Belcher, J. H. and Fogg, G. E. 1964. Chlorophyll derivatives
    and carotenoids in the sediments of two English lakes. In
    Koyam, T. and Y. Miyake, eds.; Recent Researchs in the
    Fields of Hydrosphere, Atmosphere, and Nuclear Geochemistry,
    p. 39-48, Maruzen, Tokyo.

Brown, S. R. 1969. Paleolimnological evidence from fossil
    pigments. Mitteilungen Internationale Vereingung fur
    Theoretische and Angewandte Limnologie, V. 17, p. 95-103.

Brown, S. R. and Colman, B. 1963. Oscillaxanthin in lake
    sediments. Limnology and Oceanography, V. 8, p. 233-241.

Crowl, G. H. 1980. Woodfordian age of the Wisconsin glacial
    border in northeastern Pennsylvania. Geology, V. 8, p.
    51-55.

Crowl, G. H. and Seron, W. D. 1980. Glacial border deposits of
    Late Wisconsinan Age in northeastern Pennsylvania.
    Pennsylvania Geologic Survey, General Geology Report 71, 4th ser.

Czeczuga, B. 1965. Quantitative changes in sedimentary
    chlorophyll in the bed sediment of Lake Mikolajki during
    the postglacial period; Schweizerische Zeritschrift
    Hydrologie, V. 27, p. 88-98.

Davis, M. B. 1965. Phytogeography and palynology of northeastern
    United States. In Wright and Frye, eds. Quaternary of the
    United States, p. 377-401.

-------. 1969. Climate changes in southern Connecticut recorded
    by pollen deposition at Rogers Lake. Ecology, V. 50, p.
    409-522.

-------. 1976. Pleistocene biogeography of temperate deciduous
    forests. Geoscience and Man, V. 13, p. 13-26.

-------. 1981. Mid-Holocene Hemlock Decline: Evidence for a
    Pathogen or Insect Outbreak. This Volume.

Davis, M. B. and Webb, T., III. 1975. The contemporary dis-
    tribution of pollen in eastern North America: a comparison
    with the vegetation. Quat. Research, V. 5, p. 395-434.

Deevey, E. S. 1939. Studies on Connecticut lake sediments:
    I. A. post-glacial climatic chronology for southern
    New England. American Jour. of Science, V. 237, p.
    691-724.

-------. 1949. Biogeography of the Pleistocene. Geol. Soc.
    of America Bull., V. 60, p. 1315-1416.

Denny, C. S. 1956. Surficial geology and geomorphology of
    Potter County, Pennsylvania. U.S.G.S. Prof. Paper 288,
    72 p.

Erdtmann, G. 1943. An introduction to Pollen Analysis. Ronald
    Press, New York, 239 p.

-------. 1952. Pollen morphology and plant taxonomy. Almqvist
    and Wiksell, Stockholm, 525 p.

Faegri, K. and Iverson, J. 1964. Textbook of Pollen Analysis.
    Hafner Press, New York, 296 p.

Fogg, G. E. and Belcher, J. 1961. Pigments from the bottom
    deposits of an English lake. New Phyt., V. 60, p. 129-142.

Goodlett, J. C. 1954. Vegetation adjacent to the border of the
    Wisconsin drift in Potter County, Pennsylvania. Harvard
    Forest Bulletin, V. 25, 93 p.

Gorham, E. 1960. Chlorophyll derivatives in surface muds from
    the English lakes. Limnology and Oceanography, V. 5, p.
    29-33.

-------. 1961. Chlorophyll derivatives, sulfur and carbon in
    cores from the English Lakes. Can. Jour. of Botany, V. 39,
    p. 333-338.

Gorham, E. and Sanger, J. E. 1972. Fossil pigments in the
    surface sediments of a meromictic lake. Limnology and
    Oceanography, V. 17, p. 618-622.

-------. 1975. Fossil pigments in Minnesota lake sediments and
    their bearing upon the balance between terrestrial and
    aquatic inputs to sedimentary organic matter. Verhandlungen
    Internationale Vereingigung für Limnologie, V. 19, p. 2267-
    2273.

-------. 1976. Fossil pigments as stratigraphic indicators of
    cultural eutrophication in Shagawa Lake, northeastern
    Minnesota. Geol. Soc. America Bull., V. 87, p. 1638-1642.

Griffiths, M. and Edmundson, W. T. 1975. Burial of oscillaxan-
    thin in the sediment of Lake Washington. Limnology and
    Oceanography, V. 20, p. 942-952.

Griffiths, M.,Perrott, P. S., and Edmundson, W. T. 1969.
    Oscillaxanthin in the sediment of Lake Washington.
    Limnology and Oceanography, V. 14, p. 317-326.

Klaus, W. 1978. On the taxonomic significance of tectum
    sculpture characters in alpine Pinus species. Grana,
    V. 17, p. 161-166.

Kuchler, A. W. 1964. Potential natural vegetation of the
    conterminous United Stated. Amer. Geogr. Soc. Spec.
    Publ. No. 36.

Lewis, H. C. 1884. Report on the terminal moraine in
    Pennsylvania. Pennsylvania Geol. Survey, 2nd ser.,
    Report Z.

Likens, G. E. and Davis, M. B. 1975. Post-glacial history of
    Mirror Lake and its watershed in New Hampshire, U.S.A.:
    an initial report. Verhandlungen Internationalen Verein
    Limnologie, V. 19, Part 2, p. 982-993.

McAndrews, J. H., Berti, A. A., and Norris, G. 1973. Key to the
    Quaternary pollen and spores of the Great Lakes region.
    Life Sci. Misc. Publ., R. Ont. Mus., 61 p.

Martin, P. S. 1958. Taiga-Tundra and the Full-Glacial period
    in Chester County, Pennsylvania. Am. Jour. of Science,
    V. 256, p. 470-502.

Miller, N. G. 1973. Lake-Glacial and Postglacial vegetation
    change in southwestern New York State; Bulletin 420,
    N.Y. State Museum and Science Service.

Moore, P. D. and Webb, J. A. 1978. An illustrated guide to
    pollen analysis. J. Wiley and Sons, N.Y., 133 p.

Sanger, J. E. and Crowl, G. H. 1979. Fossil pigments as a
    guide to the paleolimnology of Browns Lake, Ohio. Quat.
    Research, V. 11, p. 342-352.

Sanger, J. E. and Gorham, E. 1970. The diversity of pigments
    in lake sediments and its ecological significance.
    Limnology and Oceanography, V. 15, p. 59-69.

-------. 1972. Stratigraphy as a guide to the postglacial
    history of Kirchner Marsh, Minnesota. Limnology and
    Oceanography, V. 17, p. 840-854.

-------. 1973. A comparison of the abundance and diversity of
    fossil pigments in wetlands peats and woodland humus layers.
    Ecology, V. 54, p. 605-611.

Sirkin, L. A. 1967. Late-Pleistocene pollen stratigraphy of
    western Long Island and eastern Staten Island, New York.
    In Cushing, E. G. and H. E. Wright, eds., Quaternary
    Paleoecology, Yale University Press, p. 249-274.

Sirkin, L. A. 1976. Block Island, Rhode Island: evidence of
    fluctuation of the late Pleistocene ice margin. Geol. Soc.
    America Bulletin, V. 87, p. 574-580.

--------. 1977. Late Pleistocene vegetation and environments
    in the middle Atlantic Region. In Newman, W. S. and
    Salwen, B., ed., Amerinds and their paleoenvironments in
    northeastern North America. N.Y. Acad. Sci. Annals.,
    V. 288, p. 206-217.

Ueno, Jitsuro. 1958. Some palynologic observations of Pinacea.
    Jr. Inst. Polytechnics, Osaka City Univ., Ser. D, V. 9,
    p. 163-187.

Vallentyne, J. R. 1956. Epiphasic carotenoids in post-glacial
    lake sediments. Limnology and Oceanography, V. 1, p. 252-
    262.

--------. 1957. The molecular nature of organic matter in lakes
    and oceans, with lesser reference to sewage and terrestrial
    soils. Jr. Fisheries Research Board of Canada, V. 14,
    p. 33-82.

--------. 1960. Fossil pigments. In Allen, M. B., ed.,
    Comparative biochemistry of photoreactive systems.
    Academic Press, N.Y., p. 83-105.

Whitehead, D. R. 1979. Late-Glacial and Postglacial history of
    the Berkshires, western Massachusetts. Quat. Research,
    V. 12, p. 333-357.

Zullig, H. 1961. Die Bestimmung von Myxovanthophyll in
    Borhprofilen zum Nachweis vegangener Blaualgenentfaltung.
    Internationale Vereingung für Theoretische und Angewandte
    Limnologie, Verhandlung, V. 14, p. 263-270.

POLLEN STUDY OF BURIED TREE SITE NEAR MARQUETTE MORAINE

W. James Merry

University of Georgia

Athens, Georgia   30601

Leverett (1929) described a moraine near Marquette, Michigan as extending southeasterly from the Huron Mountains to Little Lake. Hughes (1971) was able to give more detail of this moraine finding that from Little Lake it extended eastward though not as well defined and then in a northeasterly direction so as to connect with the Munising Moraine. South and west of the Marquette Moraine there is an extensive outwash known locally as the Sands Plains.

In 1976 the Cleveland Cliffs Iron Co. started construction of a tailings basin on the western edge of the Sands Plains about ten miles southwest of Marquette and four miles east of Palmer. This involved building an earthen dam about 100 feet high and over a mile long across the eastern end of a small valley in which a small bog lake called Lake Gribben was located. The tailings basin therefore became known as the "Gribben Basin". The material for the dam was taken from within the basin. While removing sand and gravel for this construction from a borrow pit, pieces of trees were uncovered about 20 to 30 feet below the surface. These trees were in growth position and buried in silt and fine sand. The tips of intact specimens extended up into a layer of coarse gravel and had been abraded off by a rapid flow of water carrying the gravel in well defined channels. No evidence was found that the trees had been overriden by ice. Radio-carbon dates of the outer wood of these trees averaged at about 10,000 years B.P. $\pm$ 300. A number of specimens had more than 150 annual rings. Bark characteristics of all specimens examined were of spruce and the cones found in material at the base of the trees were obviously of white spruce.

117

Methods

In 1977 samples of soil were taken for pollen analysis from a bank of the borrow pit where the largest intact specimen was uncovered. Soon after the pumps, which were constantly removing water from the pit so that work was possible, were shut off and removed. The inflow of ground water then flooded the study site. The collections were made by the gelatin capsule technique described by Anne Stevenson (1968). Processing these samples was delayed by the discovery that Lake Gribben was filled with peat to a depth of over eighteen feet. Since the lake was about to be flooded with tailings, immediate coring was necessary. After making a study of these peat cores for a Master's thesis, Miss Kathy Maxson used her newly acquired ability to analyze the sample from the borrow pit.

Since they contained much sand and clay, the samples were treated in a routine manner with 10% HCl, 52% HF, 10% KOH and acetolysis as described by Faegri and Iversen (1975). They were then dehydrated and transferred to silicone oil for mounting on slides. Eucalyptus pollen tablets were added to the samples before processing so that a quantitative determination could be made. Using 22mm. square cover glasses, most of the area on three slides was counted in order to get a usable number of grains.

As the pollen diagram shows, the samples were taken at two centimeter intervals from 808 cm. below the surface to 790 cm. This included the organic layer in which the trees were rooted and about 4 cm. beneath that layer. All the roots observed spread horizontally and only a few very fine roots extended into the coarse sand and gravel below. The samples from 790 to 740 were taken at ten centimeter intervals and those above 740 were taken according to the location of dark varve-like bands in the bank.

The following description is taken from a soil profile made by Loren W. Berndt of the Soil Conservation Service of the United States Department of Agriculture. The samples at 808, 806 and 804 were from a dark brown loamy sand which was mildly alkaline. The sample at 802 was from a dark gray mucky sandy loam also mildly alkaline. The samples from 800 through 790 were from a dark grayish brown mucky fine sandy loam; 25% spruce needles and twigs and mildly alkaline. The sample at 780 was from a brown mucky fine sandy loam; 60 to 70% spruce needles and mildly alkaline. The samples from 770 through 740 were from a brown silt loam with a few scattered woody remnants and mildly alkaline. The remaining samples were from a stratified light brown very fine sand and darker brown loamy very fine sand all mildly alkaline. The darker brown layers were sampled at 720, 701, 695, 685 and 680 cm.

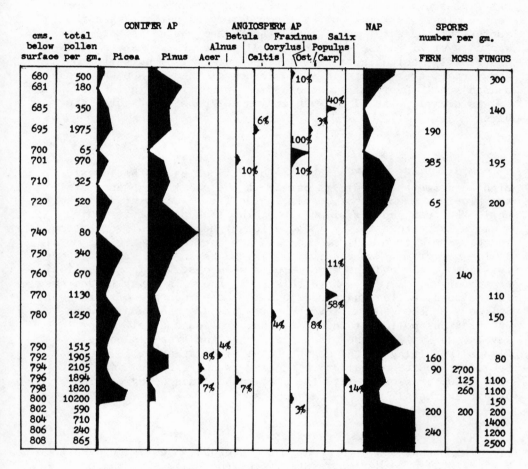

Fig. 1.   Pollen Diagram of Bank of Borrow Pit in Gribben Basin
           at Location of Largest Intact White Spruce Specimen

The samples at 710, 700 and 681 were in the light brown layers. These alternating light and dark layers had a varve-like appearance but were too irregular to be true annual varves. A number of the dark layers were too thin to be sampled. As the pollen diagram shows, the dark layers contained a greater number of pollen grains. This is probably to be expected, since all the material above 780 cm. is lacustrine and the darker bands represent a slower settling out of finer particles perhaps when the surface of the water was frozen. Above the 680 cm. sample there was a layer of coarse sand and gravel at least two meters thick and clearly indicating a rapid flow of water from north to south. Above that layer to within a few inches of the surface was cross-bedded sand. Everything above 680 cm. was strongly acid and since it was clearly glacial outwash it was not considered worth sampling.

Results

Of interest is the lack of arboreal pollen in the lowest eight centimeters. This is partly in the underlying deposit of the previous glacier but mostly in the first accumulation of soil. While the total number of pollen grains is not great it would seem to indicate a fairly long period of non-arboreal vegetation. In the sample from 880 cm. there is a great increase in the total pollen per gram. Two other areas in the basin were studied though less completely and the same results were observed in the beginning of the organic layer. Above this layer there is a marked decrease in the total pollen and a decrease in the percentage of spruce pollen. At the 700 cm. level the results are based on finding one Fraxinus pollen grain and the numbers found at all upper levels are small. Since it appears that the area was being flooded resulting in the death of the trees at about 10,000 years B.P., the source of pollen in the layers above 795 cm. may have been by water flowing into the area rather than being carried by air currents.

Conclusion

This study seems to support the proposal made by Hughes and Merry (1978) of a Marquette Stadial of the Wisconsin Glacier preceded by an interval for which the name Gribben Interstadial was suggested since the trees were exposed in what is known as the Gribben Basin. The preceding glacial advance would have been the Valders Stadial which passed the Two Creeks location about 11,850 years B.P. according to Black (1970). The recession of the Valders margin was described by Saarnisto (1974) as rapid until it halted in the Upper Peninsula of Michigan slightly before 11,000 years B.P. The halt is presumed to have caused the formation of the Marquette Moraine.

The pollen and tree data indicated at least 150 years of arboreal vegetation primarily white spruce just prior to the formation of the moraine. This would have required good drainage presumably into the Lake Superior Basin. This was ended by a readvance blocking the drainage and flooding the trees. To bring about the abrasion of the tree tops which was observed the ice margin would have had to be very close. Preceding the arboreal vegetation there was a period of at least equal duration of non-arboreal vegetation. A period of at least 300 years of plant growth does not seem consistent with a brief halt in the glacial recession but would rather indicate a recession to a considerable distance north of the present south shore of Lake Superior and a major readvance to within a mile or two of the buried trees.

I wish to thank Dr. John D. Hughes for inviting me to make this study and to assist him in the broader study of the area, the results of which will be in publication soon. Acknowledgement is gratefully made for the interest and kindness of the Cleveland Cliffs Iron Co. in permitting the study during construction of the tailings basin and for the use of equipment and manpower. I also wish to thank the Company for two research grants which made it possible to do the laboratory study in 1978 and 1979. Thanks to William and Anne Benninghoff for their encouragement, interest and advice without which I would have never attempted this study. Thanks also to Kathy Maxson for her willingness and great ability to learn a new technique largely by self-education.

References Cited

Black, Robert F. 1970. Glacial Geology of Two Creeks Forest Bed, Valderan Type Locality, and Northern Kettle Moraine State Forest: Madison, Wisconsin, Univ. of Wis. Geol. and Nat. Hist. Survey Inform. Circ. 13, 40 p.

Faegri, K. and Iversen, J. 1975. Textbook of Pollen Analysis. Hafner Press, New York. 287 p.

Hughes, John D. 1971. Post Duluth Stage Outlet from the Lake Superior Basin. Mich. Academician, v. 3, no. 4, p. 71-77.

------ and Merry, W. James. 1978. Marquette Buried Forest (850 Years Old). Abstract, AAAS Annual Meetings, Washington, D.C.

Leverett, F. 1929. Moraines and Shorelines of the Lake Superior Region. U. S. Geological Survey Professional Paper 154A U. S. Govt. Printing Off.

Saarnisto, M. 1974. The Deglaciation History of the Lake
    Superior Region and its Climatic Implications: Quaternary
    Research, v. 4, p. 316-339.

Stevenson, Anne L. 1968. A New Technique for Obtaining Uniform
    Volume Sediment Samples for Pollen Analysis. Pollen et
    Spores, v. 10, p. 463-464.

VEGETATION MAPS FOR EASTERN NORTH AMERICA: 40,000 YR B.P. TO THE PRESENT

Paul A. Delcourt, Dept. of Geol. Sci., Univ. of
Tennessee, Knoxville, Tennessee 37916 and Hazel R.
Delcourt, Environmental Sci. Div., Oak Ridge National
Lab., Oak Ridge, Tennessee 37830 and Program for
Quaternary Studies of the Southeastern United States,
Univ. of Tennessee, Knoxville, Tennessee 37916

ABSTRACT

Modern pollen assemblages produced by extant vegetation in
eastern North America provide analogs at the formation and subformation level for interpreting and mapping Quaternary vegetation
from pollen-stratigraphic evidence. Radiocarbon-dated pollen
diagrams from 100 localities across eastern Canada and the
eastern United States are the "control" points for construction
of seven paleovegetation maps spanning the past 40,000 years.
For each site, the fossil pollen spectra are compared with data
from a wide geographic array of modern pollen samples and
translated in qualitative terms into paleovegetation. The times
for which paleovegetation is mapped include 40,000 yr B.P.
(the late Altonian Subage), 25,000 yr B.P. (the Farmdalian
Subage), 18,000 yr B.P. (the peak in Late Wisconsinan continental
glaciation during the Woodfordian Subage), 14,000 yr B.P. (the
late-glacial interval during the Woodfordian Subage), 10,000 yr
B.P. (the early-Holocene interval), 5,000 yr B.P. (the mid-
Holocene interval), and 200 yr B.P. (the late-Holocene interval).

The vegetation reconstructions illustrate marked contrasts in
phytogeographic patterns between glacial and interglacial conditions as well as vegetation response to shorter-term climatic
oscillations. Spatial and temporal continuity of major vegetation

types is reflected through the late Quaternary at the formation
and subformation level of mapping. Although the broadly defined
vegetation types shift in position and through time as discrete
units, species composition and abundance has varied within each
vegetation type according to the individualistic responses of
plant populations to environmental changes. The paleovegetation
maps emphasize that more than 60 percent of the last 40,000 years
has been characterized by environmental conditions transitional
between extreme glacial and nonglacial regimes.

INTRODUCTION

. Vegetation reconstructions based upon late Quaternary plant-
fossil records from eastern North America reflect the major
climatic changes that have occurred during the last glacial/
interglacial cycle (M. Davis, 1969, 1976; Whitehead, 1973;
Wright, 1976, 1977; Watts, 1980a). The coverage of radiocarbon-
dated Quaternary sites north of 40° N latitude and east of 100° W
longitude has been adequate for qualitative reconstructions of
vegetation history of the previously glaciated landscape
(Ritchie, 1976; Wright, 1977; Webb and Bernabo, 1977). However,
the stratigraphic sections from most of these sites are limited in
time span to the last 14,000 years, i.e., the late-glacial and
postglacial periods. Near the glacial margin, sites with full-
glacial and older sedimentary sequences occur only in ice-free
pockets such as the drumlin fields of central Minnesota (Birks,
1976) or within proglacial lakes (Berti, 1975a,b).

Isochrone and isopoll maps have been published that depict
postglacial changes in distribution of individual pollen types,
in turn reflecting shifts in areal distributions of dominant taxa
(M. Davis, 1976; Bernabo and Webb, 1977). It has been postulated
that each plant taxon has migrated northward from its full-glacial
refuge at a rate commensurate with its tolerances, competitive
strategy, and dispersal capabilities (M. Davis, 1976, 1978;
Bernabo and Webb, 1977). Until recently, however, Pleistocene
and Holocene sites investigated south of 40° N in eastern North
America have been few and widely separated (Whitehead, 1973),
making vegetation reconstructions of refuge areas in the Southeast
tentative and highly speculative. In the past eight years since
Whitehead's synthesis on the Quaternary vegetation history of the
southeastern United States, numerous additional sites have been
studied in that region (Watts, 1973, 1975a, 1975b, 1979, 1980a,
b; Bryant, 1977; King and Allen, 1977; King, 1978a,b; Whitehead,
1980; Delcourt and Delcourt, 1977; H. Delcourt, 1979; P. Delcourt,
1980; Delcourt et al., 1980). With discovery and analyses of
these sites from the Southeast, it is now timely to reexamine the

spatial and temporal coverage of palynological data available
from Quaternary sites throughout all of the eastern United States
and eastern Canada.

Our objective is to use the empirical relationships of
modern pollen assemblages and extant vegetation in order to
translate fossil pollen assemblages in qualitative terms of past
vegetation. Thus, for selected times in the past, the spatial
distribution of Quaternary sites with pollen data of appropriate
age provides control points for mapping vegetation in eastern
North America at the formation and subformation level of classification. We present a time-series of paleovegetation maps extending from 200 yr B.P. back to 40,000 yr B.P. Forty thousand
yr B.P. is selected as the oldest time plane for this study, as
this date represents the effective limit for routine age determinations in many radiocarbon-dating laboratories. These maps
document the major changes in distribution of late-Quaternary
vegetation types, along with the changing geography of ice-sheet
margins, major proglacial and postglacial lakes and marine
shorelines.

METHODS

Characterization of Modern Vegetation Based Upon Pollen Assemblages

Studies of the composition of modern pollen spectra using
surface sediments from lakes, bogs, swamps, and other depositional
environments across eastern North America have shown that distinctive pollen assemblages correspond with major vegetation
types at formation and subformation level (King and Kapp, 1963;
Ritchie and Lichti-Federovich, 1967; R. Davis and Webb, 1975;
Webb and McAndrews, 1976; Richard, 1977; and Webb et al., 1978).
Qualitative reconstructions of past vegetation rest upon the
assumption that assemblages of fossil pollen grains adequately
reflect the relative abundances of vascular plant taxa in the
vegetation formerly surrounding the site of deposition (Wright,
1967; Solomon and Harrington, 1979).

R. Davis and Webb (1975 and NAPS appendices) presented pollen
spectra and characterized the modern pollen rain from major
vegetation units occurring today in eastern North America. Using
the pollen spectra from these vegetation units, we calculated
the mean percent, first standard deviation, and range in percentages for each of the principal tree pollen types based upon
the arboreal-pollen sum. Correspondingly, the total percent for
herb pollen was calculated from the upland-pollen sum of arboreal
and nonarboreal pollen (Table 1).

In modern Tundra environments in northern Canada, total herb pollen is generally greater than 25% (Table 1). Birch, alder, willow, and spruce pollen also represent major components of the pollen rain. Modern Tundra environments typically have values of pollen influx that are less than 5,000 grains/cm^2/yr (Ritchie and Lichti-Federovich, 1967).

Today within the Boreal Forest Region of central and eastern Canada, the percentage values for total herb pollen are much lower than in Tundra (Table 1). Typical arboreal-pollen assemblages from the Boreal Forest include spruce, Diploxylon pine, birch, fir, and larch as the dominants (Table 1). Pollen influx values are greater than 5,000 grains/cm^2/yr (Ritchie and Lichti-Federovich, 1967). In the paleovegetation maps, an open Spruce Forest is mapped where spruce-pollen percentages are high (greater than 20% of the arboreal-pollen sum) and Diploxylon pine values are low (less than 15%). The modern transitional zone of Boreal Forest/Tundra in northern Canada corresponds with a spruce-dominated pollen assemblage where spruce values average 52% of the tree pollen sum and herb pollen constitutes about 14% of the sum of total upland pollen (Table 1). Spruce-Jack Pine Forest is mapped where spruce-pollen percentages are high and Diploxylon pine values range between 15% and 40%. Thus, within this vegetation type, spruce is a forest dominant and jack and/or red pine a subdominant (Bernabo and Webb, 1977). Where Diploxylon pine values are greater than 40% of the tree pollen and spruce values are less than the pine, the corresponding vegetation is designated as Jack Pine-Spruce Forest. Such boreal forests today include jack and/or red pine as a forest dominant and spruce as a subdominant (Bernabo and Webb, 1977).

The Mixed Conifer-Northern Hardwoods Forest corresponds with a distinctive pollen assemblage including hemlock, Haploxylon pine, Diploxylon pine, spruce, fir, oak, birch, elm, ash, ironwood, maple, and beech (Table 1).

The Deciduous Forests of the eastern United States and southeastern Canada correspond with pollen assemblages characteristically dominated by oaks and including pollen types of many cool-temperate and warm-temperate deciduous arboreal taxa (Table 1). For this study, deciduous forests are differentiated into three areally significant forest types. The Oak-Hickory Forest has oak as the dominant arboreal pollen type and at least 2.5% of hickory pollen. Diploxylon pine values typically are less than 15%. The Mixed Mesophytic Forest is associated with the distinctive pollen assemblage including many deciduous and evergreen taxa such as oak, maple, beech, basswood, elm,

Table 1. Modern pollen assemblages of the major vegetation types in eastern North America. The pollen spectra, presented in R. Davis and Webb (1975 and NAPS appendices), are summarized with tree-pollen percentages calculated on the arboreal-pollen (AP) sum and total herb-pollen percentages based on the upland-pollen sum of arboreal and nonarboreal (NAP) pollen. The number of pollen spectra used to characterize modern pollen rain is indicated by "n".

Vegetation Type		Birch (Betula)	Alder (Alnus)	Willow (Salix)	Spruce (Picea)	Pine (Pinus)	Larch (Larix)	Fir (Abies)
Tundra								
AP data, n=5	X̄	34.0	20.0	14.0	19.6	10.9	0.2	0.0
	SD	11.2	16.6	30.8	9.6	5.5	0.5	0.0
NAP data, n=5	Range	15.0-43.5	3.0-46.0	0.0-69.0	6.0-29.0	4.0-17.0	0.0-1.0	0.0
Forest-Tundra								
AP data, n=33	X̄	19.8	11.3	0.3	52.4	13.3	0.5	1.3
	SD	11.4	7.9	0.5	15.5	7.1	0.8	3.6
NAP data, n=31	Range	4.0-51.1	0.0-33.0	0.0- 2.0	22.0-80.0	2.0-32.0	0.0-3.5	0.0-20.0
Boreal Forest								
AP data, n=56	X̄	22.3	6.3	0.8	40.3	24.5	0.2	7.4
	SD	13.5	7.2	1.4	29.1	16.6	0.4	8.8
NAP data, n=38	Range	0.0-58.0	0.0-31.0	0.0- 6.5	6.0-74.0	3.0-77.0	0.0-2.0	0.0-46.0
Mixed Conifer-Northern Hardwoods								
AP data, n=180	X̄	25.4	3.1	1.0	9.2	26.7	0.4	3.3
	SD	16.7	5.5	2.0	11.0	16.4	1.0	6.7
NAP data, n=137	Range	0.0-71.5	0.0-57.0	0.0-17.0	0.0-53.0	1.0-78.0	0.0-8.0	0.0-58.0
Deciduous Forest								
AP data, n=112	X̄	9.0	1.9	2.0	1.6	19.2	0.1	0.5
	SD	9.3	3.3	3.9	5.4	17.1	0.9	4.3
NAP data, n=84	Range	0.0-49.0	0.0-21.0	0.0-30.0	0.0-45.0	0.0-80.0	0.0-8.0	0.0-45.0
Southeastern Evergreen Forest								
AP data, n=19	X̄	1.9	2.1	0.5	0.0	44.7	0.0	0.0
	SD	3.4	4.0	0.7	0.0	18.9	0.0	0.0
NAP data, n=19	Range	0.0-12.0	0.0-16.0	0.0- 3.0	0.0	17.0-77.0	0.0	0.0

Table 1 (cont.)

Vegetation Type		Hemlock (Tsuga)	Oak (Quercus)	Hazel (Corylus)	Elm (Ulmus)	Ash (Fraxinus)	Ironwood (Carpinus-Ostrya)
Tundra							
AP data, n=5	$\bar{X}$	0.0	0.0	0.0	0.0	0.0	0.0
	SD	0.0	0.0	0.0	0.0	0.0	0.0
NAP data, n=5	Range	0.0	0.0	0.0	0.0	0.0	0.0
Forest-Tundra							
AP data, n=33	$\bar{X}$	0.2	0.4	0.2	0.2	0.1	0.0
	SD	0.4	0.6	0.3	0.4	0.3	0.1
NAP data, n=31	Range	0.0- 2.0	0.0- 1.8	0.0- 1.1	0.0- 1.5	0.0- 1.5	0.0- 0.2
Boreal Forest							
AP data, n=56	$\bar{X}$	0.2	0.6	0.1	0.1	0.1	0.1
	SD	0.5	1.0	0.4	0.4	0.2	0.2
NAP data, n=38	Range	0.0- 2.0	0.0- 3.0	0.0- 2.0	0.0- 2.0	0.0- 1.0	0.0- 1.0
Mixed Conifer-Northern Hardwoods							
AP data, n=180	$\bar{X}$	5.4	11.1	0.5	3.2	1.1	1.2
	SD	7.1	14.1	1.6	3.7	1.7	2.4
NAP data, n=137	Range	0.0-47.5	0.0-58.0	0.0-15.0	0.0-22.0	0.0-10.7	0.0-20.5
Deciduous Forest							
AP data, n=112	$\bar{X}$	3.0	35.6	0.2	3.9	2.6	1.7
	SD	5.4	18.3	0.6	4.4	3.2	6.3
NAP data, n=84	Range	0.0-22.0	0.0-85.0	0.0- 4.0	0.0-22.0	0.0-18.0	0.0-65.0
Southeastern Evergreen Forest							
AP data, n=19	$\bar{X}$	0.2	23.5	0.1	1.1	0.6	1.0
	SD	0.7	12.8	0.3	1.3	0.9	1.6
NAP data, n=19	Range	0.0- 3.0	8.0-52.0	0.0- 1.0	0.0- 5.0	0.0- 3.0	0.0- 5.0

Table 1 (cont.)

Vegetation Type		Maple (Acer)	Beech (Fagus)	Basswood (Tilia)	Poplar (Populus)	Hickory (Carya)	Walnut (Juglans)	Sycamore (Platanus)
Tundra	$\bar{X}$	0.0	0.0	0.0	0.0	0.0	0.0	0.0
AP data, n=5	SD	0.0	0.0	0.0	0.0	0.0	0.0	0.0
NAP data, n=5	Range	0.0	0.0	0.0	0.0	0.0	0.0	0.0
Forest-Tundra	$\bar{X}$	0.1	0.0	0.0	0.2	0.0	0.0	0.0
AP data, n=33	SD	0.3	0.0	0.0	0.5	0.1	0.0	0.0
NAP data, n=31	Range	0.0 -1.5	0.0	0.0	0.0-2.0	0.0- 2.0	0.0	0.0
Boreal Forest	$\bar{X}$	0.2	0.1	0.1	0.2	0.0	0.0	0.0
AP data, n=56	SD	0.4	0.3	0.6	0.7	0.2	0.1	0.0
NAP data, n=38	Range	0.0- 2.0	0.0- 1.0	0.0-4.0	0.0-5.0	0.0-1.3	0.0-1.0	0.0
Mixed Conifer-Northern Hardwoods	$\bar{X}$	2.5	3.0	0.4	0.8	0.6	0.3	0.1
AP data, n=180	SD	2.7	4.0	0.9	1.7	1.2	0.6	0.3
NAP data, n=137	Range	0.0-17.0	0.0-24.0	0.0-8.0	0.0-8.0	0.0- 6.0	0.0-4.0	0.0-2.0
Deciduous Forest	$\bar{X}$	4.5	3.9	0.3	0.8	4.4	0.8	0.4
AP data, n=112	SD	6.4	4.9	1.0	1.6	4.9	1.4	0.8
NAP data, n=84	Range	0.0-32.0	0.0-22.0	0.0-8.0	0.0-5.9	0.0-32.0	0.0-9.0	0.0-4.0
Southeastern Evergreen Forest	$\bar{X}$	1.7	0.4	0.0	0.1	3.5	0.6	0.4
AP data, n=19	SD	3.5	1.0	0.0	0.3	2.8	1.7	1.00
NAP data, n=19	Range	0.0-12.0	0.0- 4.0	0.0	0.0-1.0	0.0-10.0	0.0-6.0	0.0-4.0

P. A. DELCOURT AND H. R. DELCOURT

Table 1 (cont.)

Vegetation Type		Chestnut (Castanea)	Cedar and Juniper (Cupressineae)	Holly (Ilex-type)	Gum (Nyssa)	Sweetgum (Liquidambar)
Tundra						
AP data, n=5	X̄	0.0	0.2	0.0	0.0	0.0
NAP data, n=5	SD	0.0	0.5	0.0	0.0	0.0
	Range	0.0	0.0- 1.0	0.0	0.0	0.0
Forest-Tundra						
AP data, n=33	X̄	0.0	0.0	0.0	0.0	0.0
NAP data, n=31	SD	0.0	0.0	0.0	0.0	0.0
	Range	0.0	0.0	0.0	0.0	0.0
Boreal Forest						
AP data, n=56	X̄	0.0	0.1	0.0	0.0	0.0
NAP data, n=38	SD	0.0	0.4	0.0	0.0	0.0
	Range	0.0	0.0- 2.0	0.0	0.0	0.0
Mixed Conifer-Northern Hardwoods						
AP data, n=180	X̄	0.2	0.7	0.2	0.0	0.0
NAP data, n=137	SD	1.1	1.3	1.3	0.2	0.0
	Range	0.0-11.0	0.0- 8.7	0.0-17.0	0.0- 2.0	0.0
Deciduous Forest						
AP data, n=112	X̄	1.1	1.2	0.8	0.8	0.1
NAP data, n=84	SD	2.8	3.2	3.7	2.9	0.4
	Range	0.0-18.0	0.0-16.5	0.0-33.0	0.0-23.0	0.0- 3.0
Southeastern Evergreen Forest						
AP data, n=19	X̄	0.4	0.7	1.1	4.5	2.9
NAP data, n=19	SD	1.1	1.4	2.9	8.9	3.8
	Range	0.0- 4.0	0.0- 4.0	0.0-11.5	0.0-25.0	0.0-16.0

Table 1 (cont.)

Vegetation Type		Cypress (Taxodium)	Other Tree Types	Total Herbs (% Total NAP)
Tundra	X̄	0.0	0.0	28.2
AP data, n=5	SD	0.0	0.0	9.4
NAP data, n=5	Range	0.0	0.0	14.9-37.8
Forest-Tundra	X̄	0.0	0.0	13.8
AP data, n=33	SD	0.0	0.0	7.5
NAP data, n=31	Range	0.0	0.0	2.0-27.5
Boreal Forest	X̄	0.0	0.1	7.7
AP data, n=56	SD	0.0	0.5	7.0
NAP data, n=38	Range	0.0	0.0-3.0	1.0-37.5
Mixed Conifer-Northern Hardwoods	X̄	0.0	0.3	19.2
AP data, n=180	SD	0.0	1.8	12.2
NAP data, n=137	Range	0.0	0.0-21.0	1.0-55.6
Deciduous Forest	X̄	0.0	0.3	36.4
AP data, n=112	SD	0.0	1.2	16.5
NAP data, n=84	Range	0.0	0.0- 7.0	2.0-77.3
Southeastern Evergreen Forest	X̄	3.3	0.4	21.8
AP data, n=19	SD	5.2	0.6	12.6
NAP data, n=19	Range	0.0-18.0	0.0- 2.0	3.4-62.3

walnut, hemlock, and gum. The forest region of Oak-Chestnut
(sensu Braun, 1950) has oak as the most abundant arboreal pollen
type; in surface samples, chestnut percentages are commonly
greater than 1% of the arboreal pollen sum. Chestnut values were
much higher prior to the chestnut blight of the early 1900's
(Anderson, 1974; Solomon and Kroener, 1971; Brugam, 1978; H.
Delcourt, 1979). In the paleovegetation maps, Oak-Chestnut forests
are indicated when oak represented the predominant arboreal-
pollen type and chestnut-pollen percentages were greater than or
equal to 2%.

The arboreal pollen rain within Southeastern Evergreen
Forests is characterized by pine (primarily Diploxylon Pinus),
oak, gum, hickory, cypress, sweetgum, alder and birch (Table 1).
Southeastern Evergreen Forests are here subdivided into three
major forest types. The pollen assemblage for the Oak-Hickory-
Southern Pine Forest has high oak values (usually greater than
40%), hickory percentages greater than 2.5%, and 15% to 40%
Diploxylon pine. Today, the Southern Pine Forest in the Gulf
and Atlantic coastal plains generates a pollen rain with greater
than 40% Diploxylon pine and an associated pollen suite of oak,
hickory, sweetgum and holly (Bernabo and Webb, 1977). In the
deep South, Cypress-Gum Forest is found in coastal swamps and
river floodplains. Cypress, gum, holly, sweetgum, oak, and
cedar dominate this pollen assemblage.

The modern pollen assemblage of the Subtropical Hardwoods
Forest, documented by Riegel's 1965 study in southwestern Florida,
is dominated by mangroves, buttonwood, holly, and willow.

In addition to Tundra, other Open Vegetation types such as
Oak Savannah, Prairie, and Sand Dune Scrub are characterized by
relatively high percentages of total herb pollen. Oak Savannah
is transitional from Deciduous Forest to Prairie; the total for
nonarboreal pollen ranges from 25% up to 40% of the upland-
pollen sum. Oak, hickory, and pine are common tree-pollen types;
the herb pollen types include grass, composites, chenopods, and
other herbaceous taxa characteristic of prairies. Modern pollen
spectra from the region of Prairie are composed primarily of non-
arboreal pollen of chenopods, grass, composites including sage
and ragweed, and sedge; the average percentage for total upland-
herb pollen is 79.3% (Webb and McAndrew, 1976). Temperate
deciduous and coniferous tree taxa are also represented in the
prairie pollen assemblage. Palynological criteria for mapping
prairie include values greater than 40% herb pollen (Fig. 3;
Bernabo and Webb, 1977), presence of prairie indicators such as
Amorpha, Petalostemum, and Dalea, and an arboreal-pollen
component comprised of oak, Diploxylon and Haploxylon pine,

hickory, and elm. The Sand Dune Scrub, largely restricted to isolated favorable habitats of peninsular Florida, typically has a modern pollen rain with greater than 40% herb pollen. The Sand Dune Scrub has heliophytic herb types including rosemary (Ceratiola), spikemoss, myrtle, and Polygonella (Watts, 1975a).

Selection of Sites and Time Planes

Radiocarbon-dated pollen diagrams were selected for 100 localities across eastern North America from the many diagrams available in the literature (Fig. 1; Table 2). Where several diagrams were available from a small area, the pollen diagram with the longest temporal record and the most complete sequence of radiocarbon-dated sediments was chosen in order to ensure fairly uniform spatial and temporal coverage.

Seven time planes were selected for vegetation mapping, 40,000 yr B.P., 25,000 yr B.P., 18,000 yr B.P., 14,000 yr B.P., 10,000 yr B.P., 5,000 yr B.P., and 200 yr B.P. Vegetation reconstructions prepared for these times illustrate marked contrasts in vegetation patterns for the last glacial/interglacial cycle and for shorter term (stadial/interstadial) oscillations superimposed upon this climatic cycle. The time plane of 40,000 yr B.P. corresponds with the late Altonian Subage of the Wisconsinan Age of continental glaciation (the late Port Talbot II Interstadial as recognized by Dreimanis and Karrow, 1972) (Table 3). A relative peak in climatic amelioration occurred at approximately 25,000 yr B.P., corresponding to the midpoint of the Farmdalian Subage, which dates between 28,000 and 23,000 yr B.P. in the Midwestern and Southeastern United States (Willman and Frye, 1970; Delcourt et al., 1980). The Woodfordian Subage of the Wisconsinan Age can be subdivided into two time intervals in the Southeast: 1) the full-glacial interval from 23,000 yr B.P. to 16,500 yr B.P., and 2) the late-glacial interval from 16,500 yr B.P. to 12,500 yr B.P. (Delcourt et al., 1980). Thus, 18,000 yr B.P. coincides with the peak in Late Wisconsinan continental glaciation (Dreimanis, 1977). The map of 14,000 yr B.P. documents the character of vegetation response and deglaciation of the Laurentide Ice Sheet with the onset of limited climatic warming. The Pleistocene/Holocene boundary is time-transgressive, beginning at 12,500 yr B.P. in the Southeast and occurring as late as 11,000 yr B.P. in the Northeast. The Holocene Age can be subdivided into three time intervals with time-transgressive boundaries based on regional changes in pollen assemblages. The early-Holocene interval extends from 12,500 to 8,000 yr B.P., the mid-Holocene interval is from 8,000 to 4,000 yr B.P., and the late-Holocene interval dates from 4,000 yr B.P. to the present (Wright, 1976; H. Delcourt, 1979). Maps depicting vegetation of these three intervals were prepared for 10,000 yr B.P., 5,000 yr B.P., and 200 yr B.P.

Table 2. Availability of Pollen Data from Quaternary Sites for Selected Times

Code	Site	Citation	Ages ($\times 10^3$ yr B.P.) for which information was used for construction of maps						
			0.2	5	10	14	18	25	40
A	Albion, Quebec	Richard (1977)	X	X					
AL	Alfies Lake, Ontario	Saarnisto (1974)	X	X	X				
AP	Anderson Pond, Tenn.	Delcourt, H. (1979)	X	X	X	X	X	X	
BCP	Bartow County Ponds, Ga.	Watts (1970, 1973)	X	X	X	X	X	X	
BDP	Bog D Pond, Minn.	McAndrews (1966)	X		X				
BRC	Big Rock Creek, Ill.	Grüger, E. (1972)							X
BRL	Basswood Road Lake, New Brunswick	Mott (1975)	X	X	X				
BaB	Battaglia Bog, Ohio	Shane (1975)	X	X	X	X			
BeB	Belmont Bog, New York	Spear and Miller (1976)	X	X	X	X			
BeP	Berry Pond, Mass.	Whitehead (1979)	X	X	X				
BoB	Boriack Bog, Texas	Bryant (1977)	X	X	X	X			
BuB	Buckle's Bog, Maryland	Maxwell and Davis (1972)	X	X	X	X		X	
C	Crosswicks, N.J.	Sirken et al. (1970)	X	X	X				
C11	Camp 11, Mich.	Brubaker (1975)	X	X	X				
C59T6	Core 59T6, Fla.	Riegel (1965), Spackman et al. (1966)	X	X					
CB	Chatsworth Bog, Ill.	King (1978a, 1981)	X	X	X				
CG	Cranberry Glades, W. Va.	Watts (1979)	X	X	X				
CP	Crider's Pond, Penn.	Watts (1979)	X	X	X	X			
DFB	Disterhaft Farm Bog, Wisc.	West (1961), Baker (1970), Webb and Bryson (1972)	X	X		X			
DL	Demont Lake, Mich.	Kapp et al. (1977)	X	X	X				
DS	Dismal Swamp, Va.	Whitehead (1972)	X	X	X				
FB	Friar Branch, Tenn.	DeSelm and Brown (1978), Delcourt, P., & H. Delcourt (1979)			X				
FL	Frains Lake, Mich.	Kerfoot (1974)	X	X					

Table 2 (cont.)

Code	Site	Citation	Ages (X 10³ yr B.P.) for which information was used for construction of maps						
			0.2	5	10	14	18	25	40
G	Gabriel, Quebec	Richard (1977)	X	X	X				
GH	Garfield Heights, Ohio	Berti (1975a)		X	X			X	
GL	Green Lake, Mich.	Lawrenz (1975)						X	
GS	Goshen Springs, Ala.	Delcourt, P. (1980)	X	X	X				
HB	Hershop Bog, Texas	Larson et al. (1972)	X	X	X	X	X	X	X
HL	Hudson Lake, Ind.	Bailey (1972)	X	X	X				
HP	Hack Pond, Va.	Craig (1969)	X	X	X				
HoP	Holland Pond, Maine	R. Davis (1976), R. Davis et al. (1975)	X	X	X	X			
JL	Jacobsen Lake, Minn.	Wright and Watts (1969)	X	X	X				
K	Kenogami, Quebec	Richard (1977)	X	X	X				
KB	Kincardine Bog, Ontario	Karrow et al. (1975)	X	X	X				
KM	Kirchner Marsh, Minn.	Wright et al. (1963)	X	X	X				
L	Longswamp, Penn.	Watts (1979)	X	X	X				
LA	Lake Annie, Fla.	Watts (1975a)	X	X	X				
LI	Long Island, N.Y.	Sirken (1977)	X	X	X	X	X	X	X
LL	Lake Louise, Fla.	Watts (1971)	X	X	X				X
LLo	Lake Louis, Quebec	Vincent (1973)	X	X	X				
LMa	Lake Mary, Wisc.	Webb (1974)	X	X	X				
LMi	Lac Mimi, Quebec	Richard (1977)	X	X	X				
LWO	Lake West Okoboji, Iowa	Van Zant (1976, 1979)	X	X	X	X			
LoC	Lake of the Clouds, Minn.	Craig (1972)	X	X	X				
MBB	Mer Bleue Bog, Ontario	Mott and Camfield (1969)	X	X	X				
ML	Murtle Lake, Minn.	Janssen (1968)	X	X	X				
MM	Muscotah Marsh, Kansas	Grüger, J. (1973)	X	X	X				
MP	Mingo Pond, Tenn.	Delcourt, H. (1979)	X	X	X	X	X	X	
MS	Mont Shefford, Quebec	Richard (1977)	X	X	X				
MaL	Maplehurst Lake, Ontario	Mott and Farley-Gill (1978)	X	X	X				
MiL	Mirror Lake, N.H.	Likens and Davis (1975)	X	X	X				

Table 2 (cont.)

Code	Site	Citation	0.2	5	10	14	18	25	40
		Ages (X 10³ yr B.P.) for which information was used for construction of maps							
MoP	Moulton Pond, Maine	Bradstreet and Davis (1975)	X		X				
MuL	Mud Lake, Fla.	Watts (1971)	X	X					
NC	Nonconnah Creek, Tenn.	Delcourt, P. et al. (1980)		X		X	X	X	X
OF	Old Field, Mo.	King and Allen (1977), King (1981)				X	X	X	X
OS	Ozark Springs, Mo.	King (1973)				X	X	X	X
PB	Pittsburgh Basin, Ill.	Grüger, E. (1972)	X	X	X	X	X		
PL	Pretty Lake, Ind.	Williams (1974)	X	X	X	X	X		
PM	Pigeon Marsh, Ga.	Watts (1975b)	X	X	X	X	X		
PT	Port Talbot, Ontario	Berti (1975a)							X
PiL	Pickerel Lake, S.D.	Watts and Bright (1968)	X	X	X				
PoM	Potts Mountain, Va.	Watts (1979)	X	X	X				
PrL	Prince Lake, Ontario	Saarnisto (1974)	X	X	X				
RB	Rockyhock Bay, N.C.	Whitehead (1980)	X	X	X	X	X	X	
RL	Rogers Lake, Conn.	M. Davis (1969)	X	X	X	X			
RM	Riding Mountain, Man.	Ritchie (1964, 1976)	X	X	X				
RuL	Rutz Lake, Minn.	Waddington (1969)	X	X	X				
SB	Scarborough Bluffs, Ontario	Berti (1975a)							X
SEI	5 Southeast Iowa Sites	Baker and Hallberg (1978), Fay (1978)	X					X	X
SLN	Sud Du Lac Du Noyer, Quebec	Richard (1977)	X	X	X			X	
SP	Szabo Pond, N.J.	Watts (1979)	X	X	X				
SVB	Shady Valley Bog, Tenn.	Barclay (1957)	X	X	X	X			
SaP	Sandogardy Pond, N.H.	M. Davis (1978)	X	X	X				
SiB	Singletary Bay, N.C.	Frey (1953), Whitehead (1967)	X	X	X	X	X	X	
SiL	Silver Lake, Ohio	Ogden (1966)	X	X	X	X			
StB	Saint Benjamin, Quebec	Richard (1977)	X	X	X				
StD	Saint Davids Gorge, Ont.	Berti (1975a), Hobson and Terasmae (1969)	X						
StL	Stotzel Leis, Ohio	Shane (1976)	X		X			X	

Table 2 (cont.)

Code	Site	Citation	Ages (X 10^3 yr B.P.) for which information was used for construction of maps						
			0.2	5	10	14	18	25	40
StR	Saint Raymond de Portneuf, Quebec	Richard (1977)							
SuB	Sumner Bog, Iowa	Van Zant (1976)	X	X	X				
T	Titusville, Penn.	Berti (1975b)	X	X					X
TB	Tannersville Bog, Penn.	Watts (1979)	X	X	X				
TH	Tunica Hills, La.	P. Delcourt and H. Delcourt (1977)				X			
TL	Thane Lake, Ontario	Terasmae (1967)	X	X	X				
TP	Thompson Pond, Minn.	McAndrews (1966)	X	X	X				
TiH	Tiger Hills	Richie (1964, 1976)	X	X	X				
VB	Volo Bog, Ill.	King (1978a, 1981)	X	X	X				
VRB	Victoria Road Bog, Ont.	Terasmae (1973)	X	X	X				
VSG	Val St. Gilles, Quebec	Terasmae and Anderson (1970)							
VeB	Vestaburg Bog, Mich.	Gilliam et al. (1967)	X	X	X				
WC	Wolf Creek, Minn.	Birks (1976)	X	X	X	X			
WCB	Wigwam Creek Bog, Penn.	Sirkin (1977)	X	X	X	X	X		
WL	Wintergreen Lake, Mich.	Bailey (1979), Kapp et al. (1977)	X	X	X	X			
WMS	Warm Mineral Springs, Fla.	Sheldon (1977), pollen analyses by J. King	X	X	X				
WP	White Pond, S.C.	Watts (1980b)	X	X	X	X	X		
WS	Wedron Section, Ill.	E. Grüger (1972)						X	
WeL	Weber Lake, Minn.	Fries (1962)	X	X	X	X	—	—	—
	Number of control sites for each time plane		76	79	74	29	13	16	11

Table 3. Late-Quaternary (Geologic Time) Stratigraphy South of the Glacial
          Margin in Eastern North America

			0 yr B.P.
Holocene Age		Late-Holocene interval	
			4,000 yr B.P.
		Mid-Holocene interval	
			8,000 yr B.P.
		Early-Holocene interval	
			12,500 yr B.P.
Wisconsinan Age	Woodfordian Subage	Late-Glacial interval	
			16,500 yr B.P.
		Full-Glacial interval	
			23,000 yr B.P.
	Farmdalian Subage		
			28,000 yr B.P.
	Altonian Subage		
			>75,000 yr B.P.

Mapping Criteria

The base maps used in the time series reflect the changing geography of eastern North America during the last 40,000 years. The locations of the glacial margin, proglacial lakes, and post-glacial lakes, are in accord with the findings of Kempton and Hackett (1968), Dreimanis (1969, 1977), Prest (1970), Willman and Frye (1970), Gadd (1971), and Farrand and Eschman (1974). Shoreline positions, shifting as the result of sea level fluc-tuations, are drawn for each time plane following Field et al. (1979).

The boundaries for the vegetation types illustrated in this time series of paleovegetation maps have been established, recognizing several constraints:

(1)   the fossil pollen spectra have been interpreted in terms of vegetation at the location of each site for the time planes represented there;

(2)   the vegetation types are generalized from the site locations and within coherent physiographic regions (e.g., the occurrence of Oak-Hickory-Southern Pine forest at Goshen Springs in south-central Alabama is mapped across the sandy Gulf Coastal Plain into southeastern Texas); and

(3)   if no paleobotanical information is available for an area, vegetation data are generalized from adjacent areas with palynological data (e.g., at 25,000 yr B.P., the Cypress-Gum Forest present in the alluvial environ-ment of Nonconnah Creek, Tennessee [NC on Fig. 1] is extrapolated along the Lower Mississippi Valley).

DESCRIPTION OF THE PALEOVEGETATION MAPS

The maps presented in Figs. 2 through 9 characterize the major vegetation types and physical geography of eastern North America during the late Quaternary. The distribution of palynological sites on these maps permits the reader to visually identify the specific geographic areas and temporal intervals for which information is available. The farther back in time, the fewer and more geographically isolated are the control points and, therefore, the more speculative are the paleovegetation reconstructions.

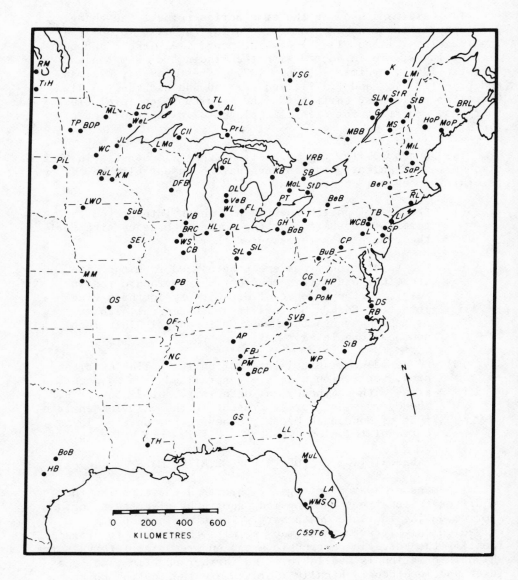

Fig. 1. Location map for sites with pollen data used in
         preparation of the late-Quaternary paleovegetation
         maps. Site names are designated by letter codes and
         are listed in Table 2.

Vegetation Map for 40,000 yr B.P. (Fig. 3)

Forty thousand years ago, during the late Altonian Subage of
the Wisconsinan Age, the Laurentide Ice Sheet extended south into
what is today the Great Lakes Region (Fig. 3). Discontinuous
pockets of Tundra flanked the southern margin of the ice; alpine
tundra has been speculatively mapped as far south as the central
Appalachians. Spruce-Jack Pine Forest occurred in the eastern
Great Lakes Region and in southeastern Canada. Jack Pine-Spruce
Forest occupied a broad latitudinal belt from the Dakotas to
Virginia. West of the Mississippi River Valley, the pollen sites
have higher percentages of total herb pollen. This trend in
percent herb pollen may represent a transition from closed
forests in the east to open, xeric forests in the west. These
would include the Jack Pine-Spruce Forest and Oak Savannah mapped
within the continental interior.

In the eastern United States, a pronounced gradient in
climate and vegetation coincided with the position of the Mixed
Conifer-Northern Hardwoods forest mapped from North Carolina,
through Tennessee, to the northern portion of the Mississippi
Embayment. Oak-Hickory-Southern Pine Forest is shown in the
southern Atlantic Coastal Plain from South Carolina to northern
Florida and in the Gulf Coastal Plain from Georgia westward into
Texas.

Deciduous taxa representative of the Mixed Hardwood Forest
may have persisted in refugial areas such as the Blufflands
(the loess-mantled uplands adjacent to the Mississippi Alluvial
Valley) as well as in dissected valleys of other major river
systems in the Southeast. Southern swamp forest of Cypress and
Gum is speculatively mapped within the poorly-drained bottomlands
of the Mississippi Alluvial Valley and along the coast of the
Gulf of Mexico. Xeric Sand Dune Scrub is shown mantling the
karst terrain of peninsular Florida. This scrub vegetation included plant cover of rosemary (Ceratiola) on active sand dunes
and isolated stands of scrub oak.

Vegetation Map for 25,000 yr B.P. (Fig. 4)

A period of mild warming occurred in eastern North America
from about 28,000 to 23,000 years ago during the Farmdalian
Subage. By 25,000 yr B.P., the Laurentide Ice Sheet had ex-
perienced a minor retreat within the western Great Lakes Region.
The Mixed Conifer-Northern Hardwoods Forest expanded to the north,
particularly along the Atlantic Seaboard. At Green Pond (one of
the Bartow County Ponds, BCP on Fig. 1) in northern Georgia, the
pollen sequence reflects a major shift in vegetation dominants

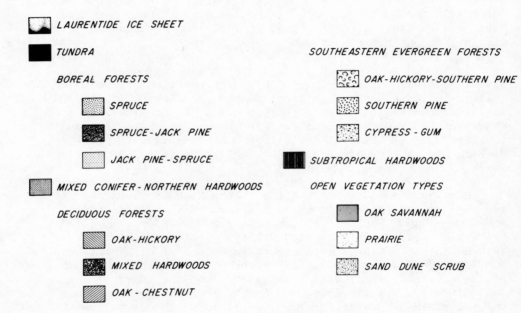

Fig. 2.   Legend for maps in Figures 3 through 9.

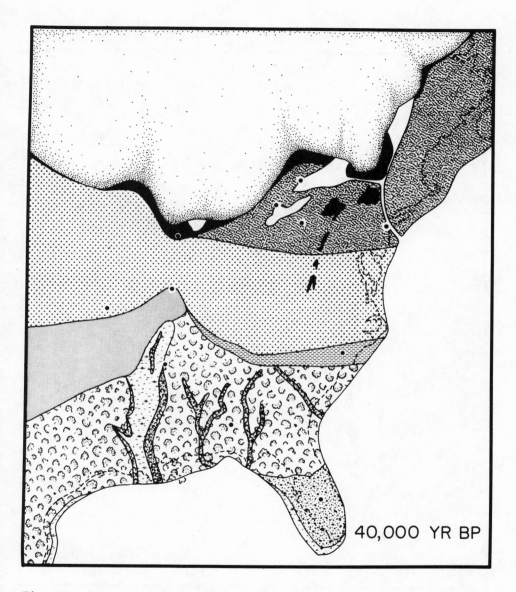

Fig. 3. Paleovegetation map for 40,000 years ago. Solid dots
locate the pollen sites that provide vegetation
information for the construction of this map.

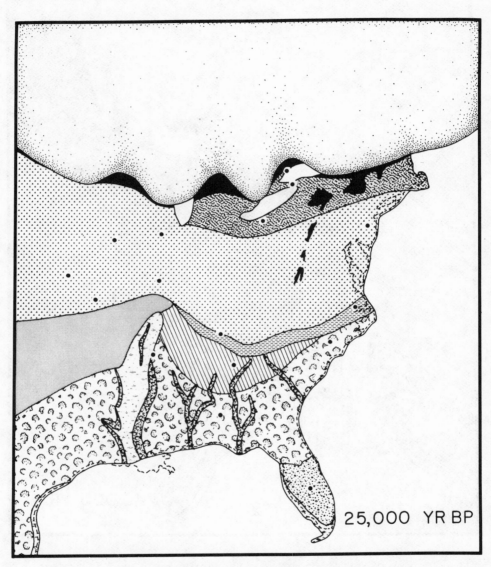

Fig. 4.   Paleovegetation map for 25,000 years ago.

from southern pine with oak and hickory to a vegetation dominated exclusively by oak and hickory (Watts, 1973). Return of
glacial meltwater to the oceans resulted in a rise in sea level
to between 10 and 5 metres below modern sea level. Consequently,
the region in Florida occupied by Sand Dune Scrub was sub-
stantially reduced.

Vegetation Map for 18,000 yr B.P. (Fig. 5)

During the peak in Late Wisconsinan Continental Glaciation
18,000 years ago, sea level was approximately 100 metres lower
than today. Fresh water was stored in continental ice sheets.
A treeless belt of Tundra 60 to 100 kilometres wide was situated
adjacent to the Laurentide Ice Sheet (Péwé, 1973; Delcourt and
Delcourt, 1979). Sorted patterned ground and block fields, i.e.,
geomorphologic features associated with alpine tundra, extended
south along the crest of the Appalachians to the Great Smoky
Mountains (Clark, 1968; Michalak, 1968). The full-glacial Boreal
Forest Region has been subdivided into three forest types. At
18,000 yr B.P., Spruce-Jack Pine Forest occupied the Great Plains
and the Ouachita and Ozark Mountains. Jack Pine Forest with sub-
dominants of spruce and fir prevailed in the Interior Low Plateaus,
midslopes of the Appalachians, and the Atlantic Coastal Plain. An
unusual ecotype of white spruce extended south in the Lower
Mississippi Alluvial Valley to the Gulf of Mexico (Delcourt and
Delcourt, 1977; Delcourt et al., 1980). White spruce and larch
were geographically isolated in the southern part of the valley,
occupying the extensive sand flats of braided streams that carried
glacial meltwater and sediment to the Gulf.

A warm-temperate forest of oak, hickory, and southern pine
persisted across the Gulf and lower Atlantic coastal plains. A
cool-temperate forest of conifers and northern hardwoods is specu-
l ively mapped from northern Mississippi to South Carolina as a
transitional zone between Jack Pine-Spruce Forest and Oak-Hickory-
Southern Pine Forest. However, no full-glacial record has yet
been found for white pine or hemlock. Based upon pollen and
plant-macrofossil documentation from Memphis, southwestern
Tennessee, Mixed Hardwood Forest is mapped along the loessial
Blufflands east of the Mississippi Valley and within dissected
valleys of major river systems on the Southeast. With a draw-
down in water table by at least 60 metres (Watts, 1975a) because
of lowered sea level, Sand Dune Scrub vegetation mantled the
landscape of the limestone platform of the Florida peninsula. No
full-glacial evidence has yet been found for cypress or gum;
thus, a belt of Cypress-Gum Forest is tentatively mapped along
the coastline of the Gulf of Mexico.

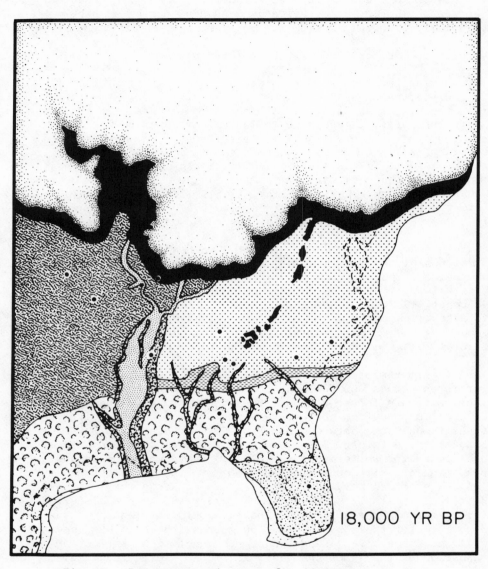

Fig. 5.   Paleovegetation map for 18,000 years ago.

Vegetation Map for 14,000 yr B.P. (Fig. 6)

At approximately 16,500 yr B.P. climatic amelioration resulted in the initial disintegration, then northward retreat, of
the Laurentide Ice Sheet (Dreimanis, 1977). During the late
glacial, the surge of meltwater was carried down the Mississippi,
diluting the salinity of ocean waters in the Gulf (Kennett and
Shackleton, 1975).

The response of vegetation to this minor climatic warming
is illustrated in the map for 14,000 yr B.P. The belt of Tundra
retreated northward with the ice, but it was separated by the
development of proglacial lakes where meltwater was dammed by
glacial moraines. Spruce Forest, corresponding to a transitional
zone of tundra and boreal forest, expanded across the freshly
deglaciated landscape. The Spruce-Jack Pine Forest expanded
eastward into Kentucky and Middle Tennessee at the expense of
the Jack Pine-dominated forest. In the mid-latitudes of the
Southeast, the cool-temperate Mixed Conifer-Northern Hardwoods
Forests expanded northward and eastward replacing Jack Pine-
Spruce Forest. The Oak-Hickory-Southern Pine Forest remained
stable in the deep South.

Vegetation Map for 10,000 yr B.P. (Fig. 7)

With major climate amelioration at 12,500 yr B.P., the Spruce
Forest within the Lower Mississippi Alluvial Valley was replaced
by Cypress-Gum Forest. By 10,000 years ago, the Laurentide Ice
Sheet had retreated from the Great Lakes Region, although major
proglacial lakes remained. Tundra persisted along the St.
Lawrence Valley and in New England. Open Spruce Forest occurred
in the Maritime Provinces of eastern Canada, New England, and
southern Manitoba. A closed Spruce-Jack Pine Forest extended from
Minnesota to southern Quebec and, immediately to the south, Jack
Pine-Spruce Forest occurred from Wisconsin to New York. With the
demise of extensive boreal forests in the Southeast, spruce and
fir populations were stranded and persisted through the Holocene
as relict "islands" at higher elevations in the southern
Appalachians. Mixed Conifer-Northern Hardwood Forest expanded
northward in advance of Oak-Hickory and Mixed Hardwood forests.
In the early Holocene, a cool, moist climate favored the widespread expansion of species-rich Mixed Hardwood Forest from 34°
to 37° N latitude in eastern North America. Prairie and Oak
Savannah developed as discrete vegetation types in the Great
Plains. Oak Savannah replaced the Sand Dune Scrub in peninsular
Florida.

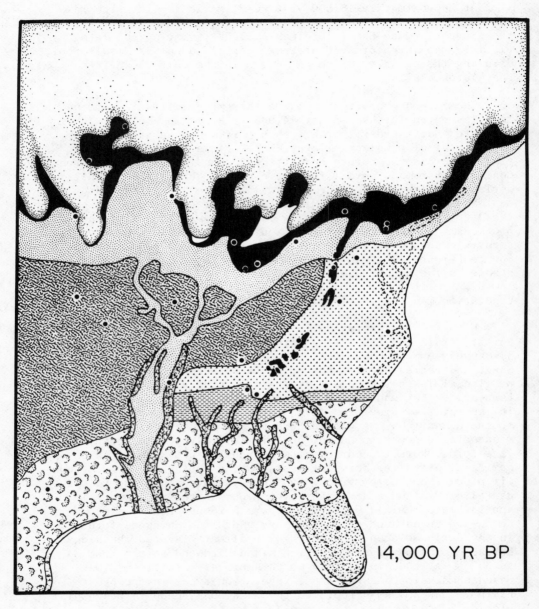

Fig. 6. Paleovegetation map for 14,000 years ago.

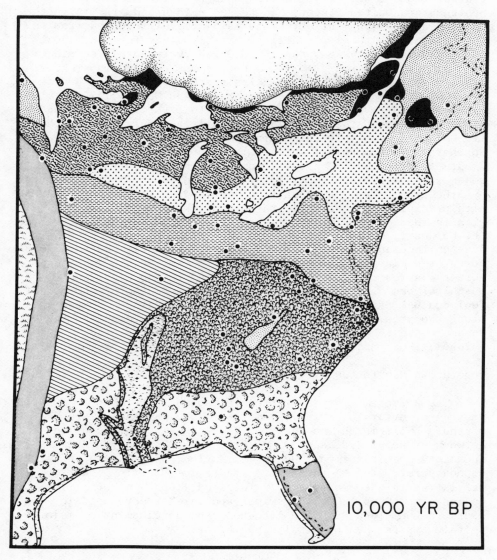

10,000 YR BP

Fig. 7. Paleovegetation map for 10,000 years ago.

Vegetation Map for 5,000 yr B.P. (Fig. 8)

The mid-Holocene interval from 8,000 to 4,000 years ago was characterized by major vegetation change in the Midwestern and Southeastern United States. Increased warmth and aridity in the Great Plains during this Hypsithermal Interval resulted from the increased strength of prevailing westerly winds (Wright, 1968, 1976). Prairie, Oak Savannah, and Oak-Hickory Forest shifted eastward as the Mixed Hardwood Forest was reduced in area. The prairie-forest boundary reached its easternmost limit approximately 7,000 years ago (Fig. 26; Bernabo and Webb, 1977). Mixed Hardwood Forest was areally restricted to favorable gorge and slope habitats in the Cumberland and Allegheny Plateaus in the mid-Holocene. As white pine migrated northward, Mixed Conifer-Northern Hardwood Forest shifted into southern Canada. Oak-Chestnut Forest became areally dominant in the central and southern Appalachians. Sea level returned to its modern position by 5,000 years ago. By that time, the modern climatic regime, with precipitation available throughout the growing season and with increased fire frequency, was established in the Southeast (P. Delcourt, 1980). Oak and hickory were replaced by southern pine on the sandy uplands of the Gulf and Atlantic Coastal Plains during the mid-Holocene interval. Oak-Hickory-Southern Pine Forest was restricted to the Piedmont and to the Ozark and Ouachita Mountains. In peninsular Flordia, Southern Pine Forest replaced Oak Savannah. Extensive swamps and marshes developed along the present coasts, and the Subtropical Hardwood Forest became established along the southernmost tip of Florida (Riegel, 1965; Spackman, et al., 1966).

Vegetation Map for 200 yr B.P. (Fig. 9)

The presettlement vegetation is mapped approximately 200 years ago, prior to extensive disruption by white settlers. A southern shift in the boundary of the Boreal Forest occurred between 5,000 yr B.P. and 200 yr B.P. due to a cooling trend and increased precipitation. During this time, the prairie/forest boundary retreated to the west, with relict Prairie pockets persisting in the predominantly forested landscapes of Illinois, Indiana, and Ohio. Coastal swamps and marshes developed in southern Louisiana with the late-Holocene development of major deltaic systems by the Mississippi River.

CONCLUSIONS

The paleovegetation maps in Figs. 2 through 9 provide a subcontinental context for evaluating vegetation response to a dynamically changing environment. Previous studies emphasized climatic extremes during the late Quaternary, e.g., the peak in glacial conditions between 23,000 and 16,500 years ago and the

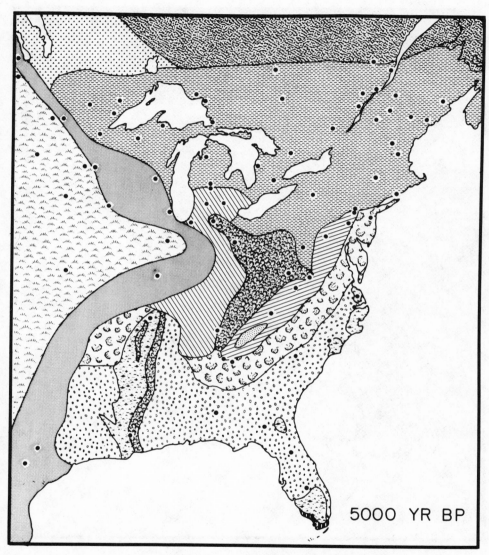

Fig. 8. Paleovegetation map for 5,000 years ago.

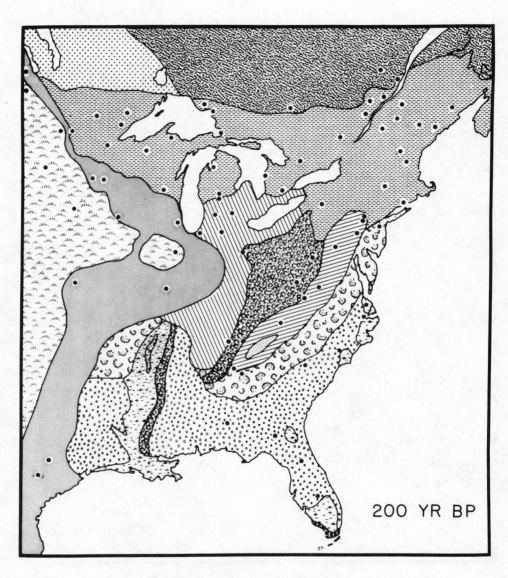

Fig. 9. Paleovegetation map for 200 years ago.

interglacial (nonglacial) conditions during the last 12,500 years. However, for over 60% of the last 40,000 years environmental conditions have been transitional between extreme glacial and nonglacial regimes. This is indicated in the broad similarity in distribution of vegetation types mapped for 40,000 yr B.P., 25,000 yr B.P., and 14,000 yr B.P.

On the formation and sub-formation level of mapping, both spatial and temporal continuity of major vegetation types are reflected in the pollen stratigraphic records available from sites throughout eastern North America. Although these small-scale maps portray the vegetation types shifting through time as discrete units, species composition and abundance has varied within each vegetation type according to the individualistic responses of plant populations to environmental changes (M. Davis, 1976; Bernabo and Webb, 1977).

During the late Quaternary, a latitudinal gradient in both magnitude and frequency of climatic oscillations has occurred between the Great Lakes Region and the Gulf Coast (Peterson et al., 1979). From North Dakota to Maine, episodes of glacial advance have been offset by ice wastage, incipient soil formation on the deglaciated terrain, and the establishment of opportunistic plant taxa that have migrated from refugial areas farther south. In contrast, the Gulf Coastal Plain has remained relatively unaffected by substantial temperature changes, even during the peak in Wisconsinan continental glaciation 18,000 years ago (Brunner and Cooley, 1976; P. Delcourt, 1980). The vegetation in the Deep South has remained relatively stable during the last 40,000 years in that an areally widespread forest mosaic of oaks, hickories, and southern Diploxylon pines has persisted in sandy upland sites. Favorable ravines and slope habitats adjacent to major river valleys across the Southeast have provided refugial areas for mesic deciduous forest taxa. These deciduous taxa, persisting during the late Quaternary in Southeastern North America, provided the seed source for the northward expansion of deciduous forests during intervals of climatic amelioration such as the Farmdalian Subage (Interstadial) and the Holocene.

This suite of maps outlines the late-Quaternary phytogeographic patterns for principal vegetation types in eastern North America. The stage is now set for development of calibration functions between modern pollen and vegetation in order to translate fossil pollen spectra in quantitative terms of past vegetation.

ACKNOWLEDGMENTS

We wish to thank the following individuals for helpful suggestions and stimulating discussions concerning paleovegetation mapping: Robert E. Bailey, Burton V. Barnes, Robert L. Burgess, Owen Davis, Stephen A. Hall, Ronald O. Kapp, John H. McAndrews, Jerry S. Olson, Allen M. Solomon, Thompson Webb, III, and Donald R. Whitehead. This research has, in part, been sponsored by the National Science Foundation's Ecology Program under Grant DEB 80-04168 and the Ecosystem Studies Program under Interagency Agreement No. DEB-77-26722 with the U.S. Department of Energy under contract W-7405-eng-26 with Union Carbide Corporation. Preparation of this manuscript was supported by the United States Heritage Conservation and Recreation Service through Contract No. C-9213 (78). This paper is Contribution Number 20 from the Program for Quaternary Studies of the Southeastern United States, Department of Geological Sciences, University of Tennessee, Knoxville, Tennessee 37916, and Publication Number 1609 from the Environmental Sciences Division, Oak Ridge National Laboratory, Oak Ridge, Tennessee 37830.

LITERATURE CITED

Anderson, T. W.   1974.   The chestnut pollen decline as a time
     horizon in lake sediments in eastern North America.
     Canadian Journal of Earth Science 11: 678-685.

Bailey, R. E.   1972.   Late- and postglacial environmental changes
     in Northwestern Indiana.   Doctoral Dissertation, Indiana
     University, Bloomington, Indiana.

--------.   1979.   Patterns of vegetational succession in the
     central Great Lakes region (Abstract).   Bulletin of the
     Ecological Society of America 60(2): 101.

Baker, R. G.   1970.   A radiocarbon-dated pollen chronology for
     Wisconsin: Disterhaft Farm Bog revisited.   Geological
     Society of America, Annual Meeting Abstract 2: 488.

--------.   1978.   Pollen analyses of Farmdalian peats in south-
     eastern Iowa (Abstract).   American Quaternary Association,
     Abstracts of the Fifth Biennial Meeting, Sept. 2-4, 1978,
     University of Alberta, Edmonton, Alberta.   p. 184.

Barclay, F. H.   1957.   The natural vegetation of Johnson County,
     Tennessee, past and present.   Ph.D. Dissertation, University
     of Tennessee, Knoxville, Tennessee.   147 pp.

Bernabo, J. C., and T. Webb, III. 1977. Changing patterns in the Holocene pollen record of northeastern North America: A mapped summary. Quaternary Research 8(1): 64-96.

Berti, A. A. 1975a. Paleobotany of Wisconsinan Interstadials, eastern Great Lakes Region, North America. Quaternary Research 5(4): 591-619.

-------. 1975b. Pollen and seed analysis of the Titusville Section (Mid-Wisconsinan), Titusville, Pennsylvania. Canadian Journal of Earth Sciences 12(9): 1675-1684.

Birks, H. J. B. 1976. Late-Wisconsinan vegetational history at Wolf Creek, central Minnesota. Ecological Monographs 46: 395-429.

Bradstreet, T. E., and R. B. Davis. 1975. Mid-Postglacial environments in New England with emphasis on Maine. Arctic Anthropology 12(2): 7-22.

Braun, E. L. 1950 (Reprinted 1974). Deciduous forests of eastern North America. Hafner Press, Macmillan Pub. Co., Inc., New York. 596 pp.

Brubaker, L. A. 1975. Postglacial forest patterns associated with till and outwash in northcentral Upper Michigan. Quaternary Research 5(4): 499-527.

Brugam, R. B. 1978. Human disturbance and the historical development of Linsley Pond. Ecology 59(1): 19-36.

Brunner, C. A., and J. F. Cooley. 1976. Circulation in the Gulf of Mexico during the last glacial maximum 18,000 yr ago. Geological Society of America Bulletin 87: 681-686.

Bryant, V. M., Jr. 1977. A 16,000 year pollen record of vegetational change in central Texas. Palynology 1: 143-156.

Clark, G. M. 1968. Sorted patterned ground: New Appalachian localities south of the glacial border. Science 161: 355-356.

Craig, A. J. 1969. Vegetational history of the Shenandoah Valley, Virginia. Geological Society of America Special Paper 123: 283-296.

-------. 1972. Pollen influx to laminated sediments: A pollen diagram from northeastern Minnesota. Ecology 53(1): 46-57.

Davis, M. B. 1969. Climatic changes in southern Connecticut
    recorded by pollen deposition at Rogers Lake. Ecology 50
    (3): 409-422.

-------. 1976. Pleistocene biogeography of temperate deciduous
    forests. Geoscience and Man 13: 13-26.

-------. 1978. Outbreaks of forest pathogens in Quaternary
    history. In Proceedings of the IV International Conference
    on Palynology, Lucknow, India, 1976-1977.

Davis, R. B. 1976. Late- and postglacial vegetational history
    of northern New England (Abstract). American Quaternary
    Association, Abstracts of the Fourth Biennial Meeting,
    October 9 and 10, 1976, Arizona State University, Tempe.
    p. 136.

Davis, R. B., and T. Webb, III. 1975. The contemporary dis-
    tribution of pollen in Eastern North America: A comparison
    with the vegetation. Quaternary Research 5: 395-434.

Davis, R. B., T. E. Bradstreet, R. Stuckenrath, Jr., and Harold
    W. Borns, Jr. 1975. Vegetation and associated environments
    during the past 14,000 years near Moulton Pond, Maine.
    Quaternary Research 5: 435-465.

Delcourt, H. R. 1979. Late Quaternary vegetation history of the
    Eastern Highland Rim and adjacent Cumberland Plateau of
    Tennessee. Ecological Monographs 49(3): 255-280.

Delcourt, P. A. 1980. Goshen Springs: Late Quaternary
    vegetation record for southern Alabama. Ecology 61(2):
    371-386.

Delcourt, P. A., and H. R. Delcourt. 1977. The Tunica Hills,
    Louisiana-Mississippi: Late glacial locality for spruce
    and deciduous forest species. Quaternary Research 7(2):
    218-237.

-------. 1979. Late Pleistocene and Holocene distributional
    history of the Deciduous Forest in the Southeastern United
    States. Veröffentilichungen des Geobotanischen Institutes
    der ETH, Stiftung Rübel (Zürich), 68: 79-107.

Delcourt, P. A., H. R. Delcourt, R. C. Brister, and L. E. Lackey.
    1980. Quaternary vegetation history of the Mississippi
    Embayment. Quaternary Research 13(1): 111-132.

DeSelm, H. R., and J. L. Brown. 1978. Fossil flora of the Friar Branch and Boyd Buchanan School sites (Abstract). Association of Southeastern Biologists Bulletin 25: 83.

Dreimanis, A. 1969. Late-Pleistocene lakes in the Ontario and the Erie basins. University of Michigan, Proceedings of the 12th Conference Great Lakes Research. pp. 170-180.

--------. 1977. Late Wisconsin glacial retreat in the Great Lakes region, North America. Annals of the New York Academy of Science 288: 70-89.

Dreimanis, A., and P. F. Karrow. 1972. Glacial history of the Great Lakes-St. Lawrence region, the classification of the Wisconsin(an) Stage, and its correlatives. International Geological Congress, Section 12: Quaternary Geology, 5-15.

Farrand, W. R., and D. F. Eschman. 1974. Glaciation of the Southern Peninsula of Michigan: A review. Papers of the Michigan Academy of Science, Arts, and Letters 7(1): 31-56.

Fay, L. P. 1978. Catalog of Quaternary palynologic and vertebrate localities in Iowa. Proceedings of the Iowa Academy of Science 85(1): 35-38.

Field, M. E., E. P. Meisburger, E. A. Stanley, and S. J. Williams. 1979. Upper Quaternary peat deposits on the Atlantic inner shelf of the United States. Geological Society of America Bulletin, Part 1, 90: 618-628.

Frey, D. G. 1953. Regional aspects of the late-glacial and postglacial pollen succesion of southeastern North Carolina. Ecological Monographs 23: 289-313.

Fries, M. 1962. Pollen profiles of late Pleistocene and recent sediments from Weber Lake, Minnesota. Ecology 43: 295-308.

Gadd, N. R. 1971. Pleistocene geology of the Central St. Lawrence Lowlands. Geological Survey of Canada, Memoir 359.

Gilliam, J. A., R. O. Kapp, and R. D. Bogue. 1967. A post-Wisconsin pollen sequence from Vestaburg Bog, Montcalm County, Michigan. Michigan Academy of Science, Arts and Letters 52: 3-17.

Grüger, E. 1972. Pollen and seed studies of Wisconsinan
    vegetation in Illinois, U.S.A. Geological Society of
    America Bulletin 83: 2715-2734.

Grüger, J. 1973. Studies on the late Quaternary vegetation
    history of northeastern Kansas. Geological Society of
    America Bulletin 84: 239-250.

Hobson, G. D., and J. Terasmae. 1969. Pleistocene geology of
    the buried St. Davids Gorge, Niagara Falls, Ontario;
    geophysical and palynological studies. Geological Survey
    of Canada, Paper 68-17.

Janssen, C. R. 1968. Myrtle Lake: A late- and postglacial
    pollen diagram from northern Minnesota. Canadian Journal
    of Botany 46(11): 1397-1408.

Kapp, R. O., R. Bailey, B. A. Manny, and A. M. Solomon
    (Organizers). 1977. Guidebook for the Paleoecology Field
    Trip in Central Lower Michigan, August 19-20, 1977. Paleo-
    ecology Section of the Ecological Society of America,
    American Institute of Biological Sciences, Annual Meeting,
    East Lansing, Michigan. 56 pp.

Karrow, P. F., T. W. Anderson, A. H. Clarke, L. D. Delorme, and
    M. R. Sreenivasa. 1975. Stratigraphy, paleontology, and
    age of Lake Algonquin sediments in southwestern Ontario,
    Canada. Quaternary Research 5(1): 49-87.

Kempton, J. P., and J. E. Hackett. 1968. Stratigraphy of the
    Woodfordian and Altonian drifts of central northern
    Illinois, pp. 27-34. In Bergstrom, R. E. (Ed.), The
    Quaternary of Illinois. College of Agriculture, University
    of Illinois, Urbana, Special Publication 14.

Kennett, J. P., and N. J. Shackleton. 1975. Laurentide Ice
    Sheet meltwater recorded in Gulf of Mexico deep-sea cores.
    Science 188 (4184): 147-150.

Kerfoot, W. C. 1974. Net accumulation rates and the history of
    cladoceran communities. Ecology 55(1): 51-61.

King, J. E. 1973. Late Pleistocene palynology and biogeo-
    graphy of the western Missouri Ozarks. Ecological
    Monographs 43(4): 539-565.

King, J. E. 1978a. Holocene pollen studies in the Prairie Peninsula (Abstract). American Quaternary Association, Abstracts of the Fifth Biennial Meeting, Sept. 2-4, 1978, University of Alberta, Edmonton, Alberta. p. 217.

-------. 1978b. New evidence on the history of the St. Francis Sunk Lands, northeastern Arkansas. Geological Society of America Bulletin 89: 1719-1722.

-------. 1981. Late Quaternary Vegetational History of Illinois. Ecological Monographs 51(1): 43-62.

King, J. E., and W. H. Allen, Jr. 1977. A Holocene vegetation record from the Mississippi River Valley, southeastern Missouri. Quaternary Research 8: 307-323.

King, J. E., and R. O. Kapp. 1963. Modern pollen rain studies in eastern Ontario. Canadian Journal of Botany 41: 243-252.

Larson, D. A., V. M. Bryant, and T. S. Patty. 1972. Pollen analysis of a central Texas bog. American Midland Naturalist 88(2): 358-367.

Lawrenz, R. W. 1975. The developmental paleoecology of Green Lake, Antrim County, Michigan. Masters Thesis, Central Michigan University, Mount Pleasant, Michigan. 76 pp.

Likens, G. E., and M. B. Davis. 1975. Post-glacial history of Mirror Lake and its watershed in New Hampshire, U.S.A.: An initial report. Verh. Internat. Verein. Limnol. 19: 982-993.

McAndrews, J. H. 1966. Postglacial history of prairie, savanna, and forest in northwestern Minnesota. Torrey Botanical Club, Memoir 22: 1-72.

Maxwell, J. A., and M. B. Davis. 1972. Pollen evidence of Pleistocene and Holocene vegetation on the Allegheny Plateau, Maryland. Quaternary Research 2(4): 506-530.

Michalek, D. D. 1968. Fanlike features and related periglacial phenomena of the Southern Blue Ridge. Ph.D. Dissertation, University of North Carolina, Chapel Hill, North Carolina. 198 pp.

Mott, R. J. 1975. Palynological studies of lake sediment profiles from southwestern New Brunswick. Canadian Journal of Earth Sciences 12(2): 273-288.

Mott, R. J., and M. Camfield. 1969. Palynological studies in the Ottawa area. Geological Survey of Canada Paper 69-38: 1-16.

Mott, R. J., and L. D. Farley-Gill. 1978. A late-Quaternary pollen profile from Woodstock, Ontario. Canadian Journal of Earth Sciences 15(7): 1101-1111.

Ogden, J. G., III. 1966. Forest history of Ohio. I. Radio-carbon dates and pollen stratigraphy of Silver Lake, Logan County, Ohio. Ohio Journal of Science 66(4): 387-400.

Peterson, G. M., T. Webb, III, J. E. Kutzbach, T. Van der Hammen, T. A. Wijmstra, and F. A. Street. 1979. The continental record of environmental conditions at 18,000 yr B.P.: An initial evaluation. Quaternary Research 12(1): 47-82.

Péwé, T. L. 1973. Ice wedge casts and past permafrost distri-bution in North America. Geoforum 15: 15-26.

Prest, V. K. 1970. Quaternary Geology of Canada (Chapter 12), pp. 676-764. In Douglas, R. J. W. (Ed.), Geology and Economic Minerals of Canada. Canadian Department of Energy, Mines, and Resources, Geological Survey of Canada, Economic Geology Report No. 1.

Richard, P. 1977. Histoire Post-Wisconsinienne de la végétation du Québec Méridional par l'analyse pollinique (Tome 1&2). Service de la Recherche, Direction Générale des Forêts, Ministère des Terres et Forêts, Gouvenement du Quebec. Tome 1, 312 pp.; Tome 2, 141 pp.

Riegel, W. L. 1965. Palynology of environments of peat formation in Southwestern Florda. Ph.D. Dissertation, Pennsylvania State University, University Park, Pennsylvania. 190 pp.

Ritchie, J. C. 1964. Contributions to the Holocene paleoecology of west-central Canada. I. The Riding Mountain area. Canadian Journal of Botany 42: 677-692.

-------. 1976. The late-Quaternary vegetational history of the Western Interior of Canada. Canadian Journal of Botany 54(15): 1793-1818.

Ritchie, J. C., and S. Lichti-Federovich. 1967. Pollen dis-persal phenomena in arctic-subarctic Canada. Review of Palaeobotany and Palynology 3: 255-266.

Saarnisto, M.   1974.   The deglaciation history of the Lake
     Superior Region and its climatic implications.   Quaternary
     Research 4(3): 316-339.

Shane, L. C. K.   1975.   Palynology and radiocarbon chronology of
     Battaglia Bog, Portage County, Ohio.   Ohio Journal of
     Science 75(2): 96-102.

-------.   1976.   Late-glacial and postglacial palynology and
     chronology of Darke County, west-central Ohio.   Ph.D.
     Dissertation, Kent State University, Kent, Ohio.   221 pp.

Sheldon, E. S.   1977.   Reconstruction of a prehistoric environment
     and its useful plants: Warm Mineral Springs (8So-19),
     Florida.   Paper presented at the Society of Economic Botany,
     June 11 to 15, 1977, Coral Gables, Florida (pollen data by
     J. E. King).

Sirkin, L. A., J. P. Owens, J. P. Minard, and M. Rubin.   1970.
     Palynology of some upper Quaternary peat samples from the
     New Jersey coastal plain.   U.S. Geological Survey Pro-
     fessional Paper 700-D: 77-87.

-------.   1977.   Late Pleistocene vegetation and environments in
     the Middle Atlantic Region.   Annals of the New York
     Academy of Sciences 288: 206-217.

Solomon, A. M., and J. B. Harrington.   1979.   Palynology models,
     pp. 338-361.   In Edmonds, R. L. (Ed.), Aerobiology, The
     Ecological Systems Approach, US/IBP Synthesis Series 10.
     Dowden, Hutchison and Ross, Inc., Stroudsburg, Pennsylvania.
     386 pp.

Solomon, A. M., and D. F. Kroener.   1971.   Suburban replacement
     of rural land uses reflected in the pollen rain of
     Northeastern New Jersey.   New Jersey Academy of Science
     Bulletin 16(1-2): 30-44.

Spackman, W., C. P. Dolsen and W. Riegel.   1966.   Phytogenic
     organic sediments and sedimentary environments in the
     Everglades-mangrove complex.   Part I: Evidence of a
     transgressing sea and its effect on environments of the
     Shark River Area of Southwestern Florida.   Palaeontographica
     Abt. B 117(4-6):   135-152.

Spear, R. W., and N. G. Miller.   1976.   A radiocarbon-dated pollen
     diagram from the Allegheny Plateau of New York State.
     Journal of the Arnold Arboretum 57(3):   369-403.

Terasmae, J. 1967. Postglacial chronology and forest history in
    the northern Lake Huron and Lake Superior Regions, pp. 45-58.
    In Cushing, E. J. and H. E. Wright, Jr. (Eds.), Quaternary
    paleoecology, Yale University Press, New Haven. 433 pp.

--------. 1973. Notes on Late Wisconsin and early Holocene
    history of vegetation in Canada. Arctic and Alpine Research
    5(3, pt. 1): 201-222.

Terasmae, J., and T. W. Anderson. 1970. Hypsithermal range
    extension of white pine (Pinus strobus L.) in Quebec, Canada.
    Canadian Journal of Earth Sciences 7(2): 406-413.

Van Zant, K. L. 1976. Late- and postglacial vegetational history
    of northern Iowa. Ph.D. Dissertation, University of Iowa,
    Iowa City, Iowa. 123 pp. (University Microfilms Inter-
    national 77-3778).

--------. 1979. Late Glacial and Postglacial pollen and plant
    macrofossils from Lake West Okoboji, Northwestern Iowa.
    Quaternary Research 12(3): 358-380.

Vincent, J. S. 1973. A playnological study for the Little Clay
    Belt, northwestern Quebec. Naturaliste Canadien 100: 59-70.

Waddington, J. C. B. 1969. A stratigraphic record of the pollen
    influx to a Lake in the Big Woods of Minnesota. Geological
    Society of America Special Paper 123: 263-281.

Watts, W. A. 1970. The full-glacial vegetation of northwestern
    Georgia. Ecology 51(1): 17-33.

--------. 1971. Postglacial and interglacial vegetation history
    of southern Georgia and central Florida. Ecology 52:
    676-690.

--------. 1973. The vegetation record of a Mid-Wisconsin Inter-
    stadial in northwest Georgia. Quaternary Research 3(2):
    257-268.

--------. 1975a. A Late Quaternary record of vegetation from
    Lake Annie, south-central Florida. Geology 3: 344-346.

--------. 1975b. Vegetation record for the last 20,000 years from
    a small marsh on Lookout Mountain, northwestern Georgia.
    Geological Society of America Bulletin 86: 287-291.

Watts, W. A. 1979. Late Quaternary vegetation of central
     Appalachia and the New Jersey Coastal Plain. Ecological
     Monographs 49(4): 427-469.

-------. 1980a. The Late Quaternary vegetation history of the
     Southeastern United States. Annual Review of Ecology and
     Systematics 11: 387-409.

-------. 1980b. Late-Quaternary vegetation history at White
     Pond on the Inner Coastal Plain of South Carolina.
     Quaternary Research 13(2): 187-199.

Watts, W. A., and R. C. Bright. 1968. Pollen, seed and mollusk
     analysis of a sediment core from Pickerel Lake, Day County,
     South Dakota. Geological Society of America Bulletin 79:
     855-876.

Webb, T., III. 1974. A vegetational history from northern
     Wisconsin: Evidence from modern and fossil pollen.
     American Midland Naturalist 92(1): 12-34.

Webb, T., III, and J. C. Bernabo. 1977. The contemporary
     distribution and Holocene stratigraphy of pollen in Eastern
     North America, pp. 130-146. In Elsik, W. C. (Ed.), Contri-
     butions of Stratigraphic Palynology, Volume 1, Cenozoic
     Palynology, American Association of Stratigraphic
     Palynologists, Contribution Series No. 5A. Houston, Texas.
     169 pp.

Webb, T., III, and R. A. Bryson. 1972. Late- and postglacial
     climatic change in the northern Midwest, USA: Quantitative
     estimates derived from fossil pollen spectra by multivariate
     statistical analysis. Quaternary Research 2(1): 70-115.

Webb, T., III, and J. H. McAndrews. 1976. Corresponding patterns
     of contemporary pollen and vegetation in Central North
     America. Geological Society of America, Memoir 145: 267-299.

Webb, T., III, G. Y. Yeracaris, and P. Richard. 1978. Mapped
     patterns in sediment samples of modern pollen from South-
     eastern Canada and Northeastern United States. Geographie
     Physique et Quaternaire 32(2): 163-176.

West, R. G. 1961. Late- and postglacial vegetational history in
     Wisconsin, particularly changes associated with the Valders
     Readvance. American Journal of Science 259: 766-783.

Whitehead, D. R. 1967. Studies of full-glacial vegetation and
     climate in southeastern United States, pp. 237-248. In
     Cushing, E. J., and H. E. Wright, Jr. (Eds.), Quaternary
     paleoecology, Yale University Press, New Haven. 433 pp.

-------. 1972. Development and environmental history of the
     Dismal Swamp. Ecological Monographs 42: 301-315.

-------. 1973. Late-Wisconsin vegetational changes in unglaciated
     eastern North America. Quaternary Research 3: 621-631.

-------. 1979. Late-Glacial and postglacial vegetational
     history of the Berkshires, Western Massachusetts.
     Quaternary Research 12(3): 333-357.

-------. 1980. Late-Pleistocene vegetational changes in north-
     eastern North Carolina. Manuscript submitted to Ecological
     Monographs.

Williams, A. S. 1974. Late-glacial-postglacial vegetational
     history of the Pretty Lake region, Northeastern Indiana.
     U.S. Geological Survey Professional Paper 686-B: 1-23.

Willman, H. G., and J. C. Frye. 1970. Pleistocene stratigraphy
     of Illinois. Illinois State Geological Survey Bulletin 94:
     1-204.

Wright, H. E., Jr. 1967. The use of surface samples in
     Quaternary pollen analysis. Review of Palaeobotany and
     Palynology 2: 321-330.

-------. 1968. History of the Prairie Peninsula, pp. 78-88.
     In Bergstrom, R. E. (Ed.), The Quaternary of Illinois.
     College of Agriculture, University of Illinois, Urbana,
     Special Report 14.

-------. 1976. The dynamic nature of Holocene vegetation, a
     problem in paleoclimatology, biogeography, and stratigraphic
     nomenclature. Quaternary Research 6(4): 581-596.

-------. 1977. Quaternary vegetation history - some comparisons
     between Europe and America. Annual Review of Earth and
     Planetary Sciences 5: 123-158.

Wright, H. E., Jr., and W. A. Watts. 1969. Glacial and
     vegetational history of northeastern Minnesota. Minnesota
     Geological Survey, St. Paul, Minnesota, SP-11. 59 pp.

Wright, H. E., Jr., T. C. Winter, and H. L. Patten. 1963. Two pollen diagrams from southeastern Minnesota: Problems in the late- and postglacial vegetational history. Geological Society of America Bulletin 74: 1371-1396.

PALEOECOLOGY OF THE CONIFERS FRENELOPSIS AND PSEUDOFRENELOPSIS (CHEIROLEPIDIACEAE) FROM THE CRETACEOUS POTOMAC GROUP OF MARYLAND AND VIRGINIA

Garland R. Upchurch* and James A. Doyle[+]

*Dept. of Botany, Univ. of Michigan
 Ann Arbor, Michigan 48109
[+]Dept. of Botany, Univ. of California
 Davis, California 95616

ABSTRACT

Proposed ecological roles for the often highly xeromorphic Mesozoic conifer family Cheirolepidiaceae, producers of Classopollis pollen, have varied widely--from coastal halophytes to upland plants. Analysis of sedimentary, megafossil, and palynological associations indicates that Pseudofrenelopsis parceramosa and Frenelopsis ramosissima in the Cretaceous Potomac Group of Maryland and Virginia were adapted to different environments. P. parceramosa (here extended from the Lower Cretaceous into the Cenomanian) occurs in gray, often bioturbated silty claystones containing few or no other identifiable plant megafossils and more or less low-diversity palynofloras characterized by Classopollis, Exesipollenites, fungi, and subordinate simple acritarchs, seen also in a subsurface interval showing evidence of marine influence. These data fit the concept that this species dominated tidally influenced coastal environments, perhaps less saline than those of P. varians in the Glen Rose Formation of Texas. In contrast, F. ramosissima occurs in different lithofacies with a high diversity of other well-preserved plant megafossils, including angiosperms, and more varied palynofloras low in fungi and lacking consistent evidence of marine plankton. It was more likely a member of richer plant communities growing in nonsaline environments, like certain xerophytes on poor soils in the modern humid tropics and subtropics. These results support the view that Cheirolepidiaceae were adapted to a wide range of habitats within the Mesozoic tropical and subtropical belts and cannot be

167

used in isolation as indicators of coastal environments or marine
influence.

INTRODUCTION

The paleoecology of the Mesozoic conifer family
Cheirolepidiaceae has been the subject of much recent speculation.
This group, which first appears in the Late Triassic, was an
important, often dominant element of Jurassic and Early
Cretaceous vegetation (Vakhrameev, 1970, 1978; Barnard, 1973).
During the rise of angiosperms in the Cretaceous, the family
declined in importance, and it became extinct by the end of the
Early Tertiary (Vakhrameev, 1970).

The Cheirolepidiaceae were unusual among conifers in at
least two respects. First, all members produced pollen of the
Classopollis (Corollina) type, characterized by a peculiar
morphology and complex ultrastructure unknown in any other group
of gymnosperms (Pettitt and Chaloner, 1964; Barnard, 1968;
Vakhrameev, 1970; Hluštík and Konzalová, 1976; Alvin et al.,
1978; Pons, 1979). Second, the vegetative shoots have a strongly
xeromorphic appearance that is often more suggestive of certain
modern xeromorphic angiosperms than familiar modern conifers.
Most Jurassic and many Cretaceous members had shoots bearing
thick, often short leaves assignable to the form genera
Brachyphyllum and Pagiophyllum (Barnard, 1968; Vakhrameev, 1970),
while in the Cretaceous genera Frenelopsis (with whorled or
opposite-decussate phyllotaxy) and Pseudofrenelopsis (with spiral
phyllotaxy), the leaves were reduced to sheathing bases, giving
the stems a jointed appearance, and the internodes performed the
function of photosynthesis (Watson, 1977; Alvin and Pais, 1978).

Ideas on the ecology of the Cheirolepidiaceae have been
based on several lines of evidence. First, the geographic distribution of Cheirolepidiaceae in relation to paleoclimatic indicators led Vakhrameev (1970, 1978) to conclude that the family
was primarily adapted to hot and/or dry climates. Cheirolepidiaceous pollen and megafossils are both most abundant at lower
latitudes, and rare or absent toward the poles (cf. also Brenner,
1976). When replotted on a world paleogeographic reconstruction
(Hughes, 1973), most Late Jurassic palynological samples from
paleolatitudes lower than about 45-55°N contain more than 50%
Classopollis, while those to the north of this belt yield only
isolated grains. Furthermore, maximum abundances of Classopollis
and cheirolepidiaceous megafossils are found both in regions and
at intervals which show evidence of aridity, such as red beds,
evaporites, and/or lowered diversity and abundance of

pteridophytes and broad-leafed gymnosperms. Examples include the
Late Jurassic of southern Eurasia (Vakhrameev, 1970, 1978), the
Early Jurassic of Western Australia (Filatoff, 1975), and the
Early Cretaceous of equatorial Africa and Brazil (Müller, 1966;
Brenner, 1976; Jardiné et al., 1974).

These distributional data are consistent with the xeromorphic
character of cheirolepidiaceous megafossils. Besides reduced
leaves, other xeromorphic features include the very thick cuticles,
deeply sunken stomata, and a succulent appearance of the stems or
leaves of many members (Jung, 1974; Watson, 1977; Pons et al.,
1980). Pons (1979) has also made an intriguing comparison
between the Cenomanian species Frenelopsis alata and the modern
North African cupressaceous conifer Tetraclinis articulata, in
which the reduced leaves have very similar marginal hairs that
serve to capture condensing atmospheric moisture during the night
in a semiarid climate.

At a more local scale, the widely noted increase in frequency
of Classopollis in transgressive marine deposits, as opposed to
continental swamp facies, has been interpreted as evidence that
Cheirolepidiaceae grew primarily in well-drained, mostly non-
depositional situations ("uplands"), rather than lowland swamps
(Chaloner and Muir, 1968; Vakhrameev, 1970; Filatoff, 1975). This
interpretation is based on the concept, first formulated by
Chaloner (1958), that a marine transgression would drown or
restrict areas occupied by lowland swamp vegetation, leaving
offshore pollen spectra dominated by plants growing in upland
areas (cf. also Chaloner and Muir, 1968, who term this phenomenon
the "Neves Effect," and Clapham, 1970).

Other authors have taken these and additional data as
evidence for an entirely different ecological interpretation--
that some, if not most, Cheirolepidiaceae were coastal halophytes,
either salt marsh shrubs or mangroves (i.e., trees growing in the
intertidal zone, with their roots in salt water). First, the
increased frequency of Classopollis in nearshore marine sediments
can be taken as evidence that its parent plants dominated coastal
areas (Pocock and Jansonius, 1961; Wall, 1965; Hughes, 1973,
1976); a modern analogy is the dominance of mangrove (Rhizophora)
pollen in nearshore marine sediments of the Orinoco delta
(Muller, 1959). Second, the fact that cheirolepidiaceous shoot
systems with highly xeromorphic features occur in floras where
the morphology of other plants suggests more mesic climates
could reflect adaptation to local saline environments. From a
morphological point of view, Frenelopsis and Pseudofrenelopsis
are particularly good candidates for halophytes: with their
reduced, sheathing leaves, thick cuticles, deeply sunken stomata,

and apparently succulent, photosynthetic stems, they are more
analogous with extant angiospermous halophytes such as Salicornia
than with modern conifers (Watson, 1977). Third, abundant, well-
preserved cheirolepidiaceous megafossils are known from several
marginal marine deposits, suggesting that the parent plants lived
near the shoreline or actually in salt water. In one case,
Brachyphyllum expansum from the Yorkshire Middle Jurassic (Harris,
1965), the remains are well preserved and abundant in sediments
where microplankton indicates marine influence, but are not known
from associated nonmarine sediments rich in other plant fossils.
In other cases, such as Pseudofrenelopsis varians from the Albian
of Texas (Daghlian and Person, 1977), Brachyphyllum nepos from the
Upper Jurassic of Germany (Jung, 1974), and Frenelopsis harrisii
from the Cenomanian of Tajikistan (Doludenko and Reymanówna, 1978),
well-preserved shoot systems are known from marine rocks, but it
is not known whether they are consistently absent from equivalent
continental facies.

    Although it seems reasonable to infer that some species of
Frenelopsis and Pseudofrenelopsis grew in saline conditions, there
are reasons to question whether they were obligate halophytes and
indicators of marine influence (as implied, for instance, by
Retallack and Dilcher, 1979). Alvin, Spicer, and Watson (1978)
caution against assuming the Cheirolepidiaceae as a whole were
ecologically uniform, noting that their great morphological
diversity and widespread dominance suggest adaptation to a wide
range of environments. In an attempt to reconstruct Wealden
plant communities by relating megafossil and pollen and spore
distributions to Allen's sedimentological models, Batten (1974,
1976) was also led to conclude that Classopollis producers
occupied a number of different habitats, including better-drained
upland and braided sandplain environments as well as coastal
areas. The remains of Pseudofrenelopsis parceramosa described by
Alvin et al. (1978) from the Barremian of the Isle of Wight in
fact occur in a lignite-rich siltsone which available evidence
indicates was deposited in freshwater, braided-stream conditions
(Alvin and Spicer, pers. comm.). Another example which supports
broad adaptations for the Cheirolepidiaceae as a whole is seen in
the distribution of Classopollis in the Lower Cretaceous rift-
valley sequences of equatorial Africa and then-adjacent Brazil
(Müller, 1966; Jardiné et al., 1974; Doyle et al., 1977, and
unpublished observations). The overwhelming dominance of
Classopollis, often occurring as well-preserved tetrads, in and
immediately below salt deposits which mark the first known
incursion of marine water into the narrow, restricted South
Atlantic rift ocean (palynozone C-IX, late Aptian: Doyle et al.,
1977) is certainly consistent with dominance of the parent
plants at the margins of saline lagoons. However, high

percentages of Classopollis are also typical of the thick underlying sedimentary sequences, which have yielded abundant sedimentological and paleontological evidence of lacustrine and
fluvial conditions (freshwater ostracodes and fishes, abundant
Botryococcus and related algal organic matter), but none of
marine influence (Reyre et al., 1966; Reyment and Tait, 1972;
Jardiné et al., 1974; Delteil et al., 1975).

PREVIOUS EVIDENCE ON POTOMAC GROUP ENVIRONMENTS

The occurrences of Frenelopsis ramosissima and Pseudofrenelopsis parceramosa reported by Fontaine (1889), Berry (1911),
and Watson (1977) from the Potomac Group of Maryland and
Virginia, the basal sedimentary unit of the Atlantic Coastal
Plain, are of special paleoecological interest, since this
sequence has long been interpreted as deposited under nonmarine
conditions. No marine animal fossils have been found in the
outcropping Potomac Group (a fact which until recently hampered
accurate dating of these sediments), although rare freshwater
clams, fishes, dinosaur bones, and insects are known (Clark et al.,
1911; our observations), and plant megafossils are locally
abundant, diverse, and well preserved. Potomac Group sediments,
including clays, cross-bedded sands, and conglomerates, show the
marked vertical and lateral heterogeneity and frequently varie-
gated colors typical of river-laid, continental deposits.
Although Fontaine (1889) believed the Potomac Group was deposited
in estuarine environments, more recent sedimentological studies
have interpreted these sediments as fluvial in origin on the basis
of sedimentary structures, stratification sequences (predominance
of fining-upward cycles), inferred current directions, geometry
of sediment bodies, and the predominance of kaolinitic and
montmorillonitic clays (Glaser, 1969; Owens, 1969; Hickey and
Doyle, 1977; Drake et al., 1979; Reinhardt et al., 1980). In the
most detailed of these studies, Glaser (1969) recognized a trend
from generally coarser and more arkosic beds in the lower Potomac,
suggesting prevalence of braided streams and more rugged source
terrains, to finer and more mature sands and more overbank
deposits in the upper Potomac, suggesting lower-gradient,
meandering streams and a more deeply weathered source area. He
noted that the Potomac Group contrasts with the overlying Magothy
Formation (Upper Cretaceous, Santonian) in lacking evidence of
tidal deposition, such as current reversals in sand bodies and
alternating thin beds of clay and pure sand. Glaser did report
a glauconitic bed attributed to the upper Potomac Group south
of Aquia Creek, Virginia as evidence that marine conditions
reached the outcrop area near the end of the Potomac deposition,
but field examination shows that this bed actually represents an

atypical, cross-bedded facies at the base of the marine Paleocene
Aquia Formation, which unconformably overlies the Potomac Group
in this area (J. P. Owens, pers. comm.; Doyle, pers. obs.).

While published data are either indicative of or consistent
with continental, nonmarine conditions at the present outcrop
belt of the Potomac Group, there are ample indications of increasing marine influence in the subsurface, toward the Atlantic
Ocean (cf. Glaser, 1969). Marine conditions were first recognized
in deep wells drilled on the Eastern Shore of Chesapeake Bay,
some 110-140 km downdip from the outcrop belt, where Potomac
Group equivalents thicken to over 1000 m (Anderson, 1948). Recent
palynological studies, supplemented by electric-log correlations
between wells, allow these wells to be correlated more accurately
with the outcrop sequence and the standard stages of the
Cretaceous (Doyle and Robbins, 1977). Anderson (1948) reported
glauconite at horizons which probably fall within palynozone I
(Barremian?-Aptian) in the Hammond well, dwarfed mollusks believed
to indicate brackish conditions in probable equivalents of upper
Zone I (Aptian) in the Esso No. 1 well, and other mollusks at
younger horizons (upper Subzone II-B and Zone III, late Albian
and early Cenomanian) in the Hammond well. In their study of
the Oak Grove well, ca. 40 km downdip from the outcrop belt in
Virginia, Reinhardt, Christopher, and Owens (1980) interpreted
the relatively coarse lower part of the Potomac Group (Zone I) as
deposited in alluvial-delta plain conditions not far from the sea,
using such evidence as the presence of glauconite (up to 10% of
the heavy mineral fraction), a pyritized foraminiferal test,
higher proportions of illite in clays, and abundant clay clast
conglomerates. In contrast, the upper part of the Potomac Group
in the same well (Zone II, Albian) shows fining-upward cycles, a
lack of glauconite, lower illite content, and locally mottled
red clays, all typical of fluvial deposits, suggesting seaward
progradation of a delta lobe over this area.

Palynological investigations on the Potomac Group have provided a similar picture of predominantly continental conditions
at the outcrop area and marine influence in the subsurface.
Mesozoic offshore marine sediments typically yield diverse and
abundant dinoflagellates and acritarchs (organic-walled microfossils of uncertain affinities, most of them probably cysts of
planktonic algae), while in presumed nearshore and lagoonal facies
lower diversity, a tendency for monodominance, fewer dinoflagellates, and a larger proportion of spiny over polygonal acritarchs
have been noted (Wall, 1965). In his monographic study of
palynology of the outcropping and shallow subsurface Potomac
Group, Brenner (1963) reported a great diversity of land plant
spores and pollen grains, but no dinoflagellates or other marine

microfossils. To the pollen and spores may be added a variety of
fungal (ascomycete) remains, which are abundant in some samples
(cf. Pirozynski and Weresub, 1979). However, subsequent work has
revealed marine microplankton in subsurface equivalents of the
Potomac Group. Dinoflagellates and smooth and spiny acritarchs
are common in many samples from deep wells on the Eastern Shore
(including the Hammond and Bethards wells mentioned above, plus
the newer Taylor well), especially from presumed late Albian and
early Cenomanian horizons (Subzone II-B and Zone III), suggesting
transgressive tendencies toward the end of the Early Cretaceous
(Doyle and Robbins, 1977). Small, fine-spined and smooth, glassy
acritarchs also occur in samples from upper Subzone II-B and
Zone III in two shallower wells drilled through the Potomac Group
near Delaware City, Delaware; significantly, they are more common
and include larger and more diverse spiny forms in the more southeasterly (downdip) of the two wells (Doyle, 1977, and unpublished).
In contrast, extensive palynological sampling of the outcrop
Potomac Group has revealed occasional small, smooth-walled
acritarchs, but only one occurrence of dinoflagellates--in a thin
gray clay bed in a predominantly red clay section near Bladensburg,
Maryland (sample 65-1 of Doyle, 1969), which interestingly
contains abundant taxodiaceous conifer twigs (Sphenolepis
kurriana sensu Berry, 1911) and ca. 80% taxodiaceous-type pollen
(Inaperturopollenites dubius), but no cheirolepidiaceous megafossils.

     These data suggested to us that the presence or absence,
abundance, and diversity of dinoflagellates and acritarchs might
provide clues on the degree of marine influence at old and new
Potomac Group localities yielding Frenelopsis and Pseudofrenelopsis. However, although it is clear that assemblages with
dinoflagellates and abundant smooth and spiny acritarchs represent the most marine end of the Potomac plankton spectrum, we have
found that we can only speculate on the salinity levels represented by samples containing only occasional smooth acritarchs.
The abundance of smooth acritarchs in samples with dinoflagellates
suggests that many such forms are marine plankton, but their
morphology is so generalized that others of them could be cysts
of freshwater algae or even fungal spores.* (Our somewhat
arbitrary criteria for separating relatively featureless
acritarchs from one-celled fungal spores in the same samples
include lack of pigmentation, glassy appearance, and a lack of
evident polarity, pores, or attachment scars.) Furthermore,
below a certain level of abundance, isolated occurrences of
plankton could be the result of landward transport by storms or
tidal currents too weak to have a significant effect on land

---

*See note added in proof, p. 196.

vegetation. Despite the fact that our study has raised as many
questions as it answers, we believe our data and analysis are of
interest in helping to set existing hypotheses on frenelopsid
ecology in perspective, in pointing out where new data are needed,
and in underlining the need for caution in drawing conclusions on
the ecology of extinct plants preserved in such highly complex,
transitional environments.

MATERIALS AND METHODS

During field work on the Potomac Group in the spring and
summer of 1976, we collected sterile shoots of Pseudofrenelopsis
parceramosa at three localities: the south side of Dutch Gap
Canal (76-27) and Drewrys Bluff (76-21; station 26 of Brenner,
1963) on the James River south of Richmond, Virginia, and Bodkin
Point (76-8; 68-65 of Doyle, 1969) on Chesapeake Bay southeast of
Baltimore, Maryland. The fossils were identified on the basis of
both external morphology and cuticular anatomy, using the taxonomic scheme of Watson (1977). We have collected no new material
of Frenelopsis ramosissima; our observations on this species are
based both on the numerous published photographs (Berry, 1911;
Watson, 1977) of material from the now defunct Fredericksburg,
Virginia and Baltimore, Maryland localities, and on Fredericksburg material assignable to this species on external morphology
in the Harvard University paleobotanical collections.

Matrix from the new P. parceramosa localities was prepared
for palynological analysis using standard methods. Observations
on the Fredericksburg (71-21) and Baltimore (71-6) localities are
based on preparations of matrix from collections at the Smithsonian Institution made for previous studies of Potomac Group
angiosperms (Doyle and Hickey, 1976; Hickey and Doyle, 1977).
Several other samples have been particularly useful for comparative purposes: a series of samples collected in November, 1979
from localities yielding Pseudofrenelopsis varians in the marine
Glen Rose Formation (early Albian?) of Somervell County, Texas
during a field trip of the American Association of Stratigraphic
Palynologists (Perkins and Langston, 1979), and core samples
kindly provided by R. A. Christopher, U.S.G.S., from the Oak
Grove, Virginia well previously studied by Reinhardt et al.
(1980). The palynostratigraphic zonation and palynological
taxonomy employed follow Brenner (1963) and Doyle and Robbins
(1977). Frequencies of pollen and spore types (Table 1) are
based on a count of 200 specimens; frequencies of plankton are
based on this number plus that of plankton encountered during
the same count.

RESULTS

Pseudofrenelopsis parceramosa--All three new Potomac localities contain shoot remains consisting of two or more connected internodes which are assignable to P. parceramosa on the basis of one leaf per node, long internodes with a 20-40 µm thick cuticle, and deeply sunken stomata with one ring of four to six papillate subsidiary cells (Figs. 1a-c). In all cases, the shoots lie parallel to the plane of bedding in gray, silty claystones, indicating at least some transport, but show no evidence of corrosion or mechanical abrasion. Associated plant megafossil remains are either subordinate, fragmentary, and unidentifiable or low in systematic diversity.

The Drewrys Bluff fossils are from a medium gray, angular clay boulder (ca. 25 x 50 cm) imbedded about 1.5 m above high tide level in a gravel unit which overlies and truncates two other cross-bedded sand and gravel units and fines upward into sands. These beds are typical of lower Potomac Group sediments described by Glaser (1969) and considered of fluvial (possibly braided-stream) origin. The clay itself is nearly massive: it shows only occasional obscure laminations and is highly mottled with small (a few mm long), thin, discontinuous lensoidal blebs and occasional vertical streaks of slightly lighter, siltier material, indicating biological disturbance by roots or burrowing animals prior to compaction. Some bedding planes exhibit fine, lighter-colored, branching markings (Fig. 1d); we have not determined whether these are of plant or animal origin (e.g., polychaete feeding structures: cf. Basan, 1978). The P. parceramosa specimens consist of isolated internodes and several unbranched shoots with two to four internodes each, indicating differing degrees of fragmentation during or after transport (Fig. 1c). The only associated macroscopic plant remains are polygonal chunks (up to 1 cm) of fusainized wood (charcoal) and a few unrecognizable small leaf fragments. Higher beds at Drewrys Bluff yield more diverse floras, including angiosperm leaves (Doyle and Hickey, 1976; Upchurch, 1978).

The palynoflora from the Drewrys Bluff clayball is readily assignable to the lower part of Zone I (Barremian or early Aptian?). As noted by Brenner (1963), Zone I is defined less on the presence of characteristic species than on the absence of species which appear at the base of Zone II (early to late Albian?). Most spores and pollen grains from Drewrys Bluff are in fact long-ranging forms which occur in both zones, such as Cyathidites minor, several Cicatricosisporites and Appendicisporites species, Aequitriradites (Cirratriradites)

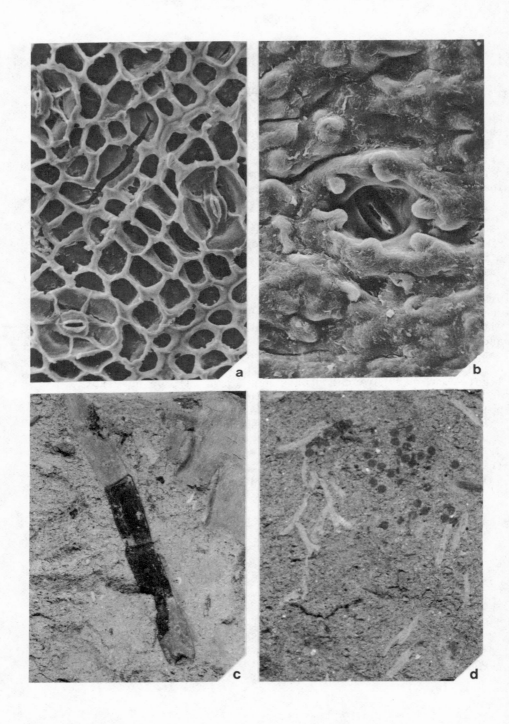

Fig. 1a,b: Scanning electron micrographs of Pseudofrenelopsis parceramosa cuticle, Dutch Gap, same specimen. a: Inner surface of cuticle, showing 4-6 subsidiary cells per stomate and thick cuticular flanges between epidermal cells, X 360. b: Outer surface, showing sunken stomate overarched by papillate subsidiary cells, and papillate non-subsidiary cells (not papillate in all specimens), X 900.

Fig. 1c: Shoot fragment of P. parceramosa, Drewrys Bluff, showing single leaf per node, X 2.

Fig. 1d: Bedding plane with branching trace fossils and associated round structures, Drewrys Bluff, X 4.

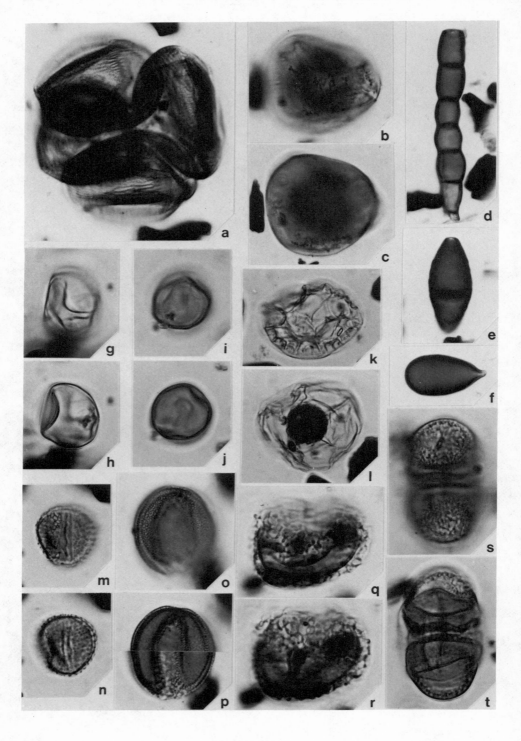

Fig. 2. Pollen, fungi, and acritarchs from the Drewrys Bluff *Pseudofrenelopsis parceramosa* locality (76-21). All light micrographs, X 1000. Coordinates: UCD Zeiss microscope UC 78 30 05524.

a: *Classopollis* tetrad, slide 76-21-1a, 16.6 X 93.1

b, c: *Exesipollenites tumulus*, two focal levels, 76-21-1a, 19.7 X 100.3

d: multicellular fungal spore, 76-21-1a, 12.8 X 103.6

e: bicellular fungal spore, 76-21-1a, 5.9 X 112.2

f: unicellular fungal spore, 76-21-1a, 15.5 X 103.7

g, h: smooth, spherical acritarch, 76-21-1a, 3.7 X 92.5

i, j: somewhat thicker-walled smooth acritarch, 76-21-1a, 3.6 X 107.8

k: wrinkled-reticulate acritarch, 76-21-1a, 16.0 X 101.0

l: thin-walled acritarch with central body, 76-21-1a, 12.9 X 94.0

m, n: cf. *Clavatipollenites minutus*, 76-21-1a, 19.5 X 119.1

o, p: cf. *Clavatipollenites hughesii*, 76-21-1a, 19.5 X 110.0

q, r: *Retimonocolpites* sp., 76-21-1a, 17.6 X 98.1

s, t: aff. *Decussosporites* (bisaccate), 76-21-1a, 16.0 X 93.2

spinulosus, Apiculatisporis asymmetricus, Cingulatisporites spp.,
Converrucosisporites proxigranulatus, Densoisporites perinatus,
Granulatisporites dailyi, Klukisporites, Lycopodiacidites
triangularis, L. intraverrucatus, Pilosisporites trichopapillosus,
Triporoletes, Inaperturopollenites pseudoreticulatus, Classopollis,
Exesipollenites tumulus, several bisaccate conifer types,
Inaperturopollenites dubius, Laricoidites, Callialasporites or
Araucariacites, small and medium-sized Ephedripites, and several
smooth monosulcates and small Eucommiidites (e.g., Monosulcites
glottus Brenner). However, two species are present which are
restricted to Zone I: Kuylisporites lunaris, and a small bisaccate believed to be related to Decussosporites Brenner (Figs.
2s,t), which has been proposed as a guide form for the lower part
of Zone I (Doyle and Hickey, 1976). As is typical of lower Zone
I, angiosperm pollen is rare and consists entirely of finely to
coarsely reticulate, columellar monosulcates (cf. Clavatipollenites hughesii, cf. C. minutus, Retimonocolpites spp.: Figs.
2m-r). We have not observed the very coarsely reticulate species
R. peroreticulatus, which appears just above the Barremian-Aptian
boundary in England (Hughes et al., 1979) and is almost ubiquitous
after its appearance in the Potomac Group; this raises the possibility of a Barremian rather than Aptian age, slightly older than
Dutch Gap (cf. below).

For paleoecological purposes, the most striking feature of
the Drewrys Bluff sample is the overwhelming dominance of
Classopollis (78%), often preserved as permanent tetrads (Fig.
2a). This contrasts with an average frequency of 16% in 20 Zone
I samples counted by Brenner (1963), and is in excellent agreement with the megafossil evidence for local dominance of P.
parceramosa, a Classopollis producer (Alvin et al., 1978). The
next most common element (10.5%) is Exesipollenites tumulus
(Figs. 2b,c), a monoporate form which has been compared (rather
unconvincingly) with modern Taxodiaceae because of its single
pore, but which is more similar to pollen recently extracted from
the Jurassic bennettitalian fructification Williamsoniella
lignieri (Harris, 1974), and which has been noted as an associate
of Classopollis in sequences as far removed as the Lower Jurassic
of Australia (Filatoff, 1975) and the Potomac Group (Brenner,
1963). The pollen and spores are associated with rather monoto-
nous brown wood (vitrain) and frenelopsid cuticle fragments, a
relatively high abundance and diversity of unicellular and
multicellular fungal remains (Fig. 2d-f), some of which resemble
forms figured by Pirozynski and Weresub (1979) and compared with
modern marine-brackish taxa, and sporadic acritarchs (0.5%).
Most of the latter are small, smooth, glassy spheromorphs (Figs.
2g-j), but we have also observed wrinkled-reticulate spheres

(Fig. 2k), thin envelopes with a small central body (Fig. 21), and more problematical larger fragments.

Samples from the Glen Rose Formation of Texas are of paleo-ecological interest for both similarities to and differences from the Drewrys Bluff sample. The sample studied in greatest detail (79-26, Cedar Brake Camp) is from the lower part of unit 8 of Nagle (1968) and Perkins and Langston (1979), a 60 cm dolomitic mudstone bed containing Pseudofrenelopsis varians, interpreted as deposited in a tidal marsh environment, which is separated from an overlying sandy, dolomitic, calcarenite unit by a mat of P. varians remains. Although this sample is roughly a stage younger than Drewrys Bluff (early Albian?) and contains different stratigraphi-cally important minor elements, including very rare tricolpate dicot pollen, it exhibits nearly identical high frequencies of Classopollis (78%, often in tetrads) and Exesipollenites (10%), and many similar spores, gymnosperm pollen grains, and fungi. However, the planktonic element is far more conspicuous than at Drewrys Bluff, representing 23% of the terrestrial-plankton sum, and including not only smooth and spiny acritarchs but also even more abundant dinoflagellates and foraminiferal chamber linings. As at Drewrys Bluff, the association of Classopollis dominance with well-preserved Pseudofrenelopsis shoots certainly supports the concept that these conifers formed nearly monospecific stands in certain local environments. In the Glen Rose case, at least, high percentages of Classopollis in marginal marine deposits do not reflect the contribution of distant upland vegetation (Neves Effect of Chaloner and Muir, 1968). However, while the Glen Rose data therefore support the concept that P. varians was a coastal halophyte (Watson, 1977; Daghlian and Person, 1977), the markedly lower abundance and diversity of probable planktonic forms at Drewrys Bluff indicates it would be unwarranted to conclude that P. varians and P. parceramosa were both adapted to the same levels of salinity. Of course, other variables, such as climate and relative input of clastic material, preclude assuming that the differences in plankton are a simple function of differences in salinity at the two localities.

A closer and equally instructive analogy may be drawn between Drewrys Bluff and a core sample of slightly better laminated, mottled, medium gray silty clay from near the base of the Oak Grove well (OG 1323, depth 1321-24.5'), above a sample where Reinhardt et al. (1980) previously noted dominance of Classopollis and Exesipollenites (1348'). Although there is a higher contribution of inaperturate (15%) and miscellaneous gymnosperm pollen than at Drewrys Bluff, Classopollis is about four times more abundant than any other element (60.5%), and

Fig. 3a: View of exposure of Potomac Group sands and clays on south side of Dutch Gap Canal, James River, Virginia, looking west. Bed containing Pseudofrenelopsis parceramosa is exposed on beach at foot of bluff at low tide.

Fig. 3b: Branching shoot of P. parceramosa, Dutch Gap, X 1.

Fig. 3c: Shoot fragments and isolated internodes of P. parceramosa, Bodkin Point, X 1.

Exesipollenites is next most common (15.5%). Equally important
for our analogy, the pollen slides contain cuticle fragments
comparable to P. parceramosa. The tentative assignment of this
interval to Zone I by Reinhardt et al. (1980) is confirmed by the
presence of angiospermous monosulcates, including Retimonocolpites
peroreticulatus (indicating a post-Barremian age), and aff.
Decussosporites (indicating lower Zone I). Fungi are somewhat
less abundant and include fewer multicellular forms than at
Drewrys Bluff. As at Drewrys Bluff but in contrast to the Glen
Rose, possible marine plankton makes up only a small fraction of
the total flora (2%) and consists largely of small, smooth
acritarchs.

As noted above, Reinhardt et al. (1980) argued that the lower
Zone I interval at Oak Grove was deposited in an alluvial-deltaic
transition zone, not far from the shore. Their strongest evidence
for proximity of marine conditions was the presence of glauconite:
since this mineral is believed to be formed only in marine
environments, its presence (unless the result of reworking from
older rocks, not plausible here) indicates either marine conditions at the site of deposition or upstream transport by tidal
currents; i.e., significant marine influence. Unfortunately, since
Reinhardt et al. (1980) explicitly record glauconite only at
somewhat higher levels (e.g., 1222.5'), this inference cannot be
extrapolated to the OG 1323 core and hence to Drewrys Bluff without
making weakening assumptions. However, in view of the fact that
Reinhardt et al. infer similar depositional conditions throughout
the lower part of the Oak Grove section, the marked similarity in
palynofacies and lithology between OG 1323 and Drewrys Bluff,
and the presence of abundant P. parceramosa shoots at Drewrys
Bluff and similar cuticles in OG 1323, the data are certainly
consistent with the idea that these conifers dominated coastal
environments subjected to some degree of tidal influence, though
perhaps lower levels of salinity than in Texas. Conversely,
these comparisons make it quite unnecessary to interpret
Classopollis dominance at Oak Grove in terms of an upland source,
a concept tentatively supported by Reinhardt et al. (1980).

The Dutch Gap Pseudofrenelopsis parceramosa material was
collected from the lowest bed (exposed only at low tide) of a
complex, 3-4 m thick sequence of interbedded clays and cross-
bedded sands (Fig. 3a). There is a strong tendency for higher
proportions of sand and thinning and truncation of clays beds
by channel sands toward the east end of this exposure. We
tentatively interpret these beds as crevasse-splay and backswamp
deposits; however, it would be difficult to rule out a tidal-
channel origin for some units, although we have seen none of the

flaser bedding, ripples, strong bioturbation, or invertebrates
often associated with tidal conditions. The <u>P</u>. <u>parceramosa</u> bed
consists of 5-10 cm of nearly massive dark gray, silty clay with
a planar fabric but no evident lamination except near its contact
with the underlying sand; its darker color, apparently lower silt
content, more abundant lignitic plant fragments, and lack of
evident bioturbation suggest more stagnant, reducing conditions
than at Drewrys Bluff. Plant fragments do, however, show a
tendency for parallel alignment, indicating some current activity.
The <u>P</u>. <u>parceramosa</u> remains include two shoots with many internodes
each, one of them showing axillary branching (Fig. 3b). The only
other identifiable plant megafossils are frond fragments of the
bennettitalian species <u>Zamites</u> (<u>Dioonites</u>) <u>buchianus</u> (Fontaine)
Seward, but these are more common than <u>P</u>. <u>parceramosa</u> itself, and
pieces of fusainized wood and fine plant debris are more abundant.
Other beds higher in the section contain more or less different
floras, some with <u>Z</u>. <u>buchianus</u> and/or diverse fern and non-
cheirolepidiaceous conifer remains (e.g., Taxodiaceae,
<u>Podozamites</u>), and one with rare dicot leaves (Hickey and Doyle,
1977, Fig. 17).

The Dutch Gap palynoflora is readily assignable to Zone I on
the predominance of long-ranging spores and gymnosperm pollen,
presence of <u>Kuylisporites lunaris</u>, lack of Zone II index species,
and exclusively monosulcate angiosperm pollen. The presence of
aff. <u>Decussosporites</u> points to lower Zone I, while the presence of
<u>Retimonocolpites peroreticulatus</u> (Figs. 4a,b) indicates a post-
Barremian age, possibly younger than the Drewrys Bluff clayball.
Quantitatively, <u>Classopollis</u> is the most important element (21%),
more abundant than the 16% Zone I average but far less so than
in the Drewrys Bluff, Oak Grove, and Glen Rose samples, while its
familiar associate <u>Exesipollenites</u> is nearly as common (16.5%).
In further contrast to Drewrys Bluff, bisaccate and probable
taxodiaceous conifer pollen types are both almost as abundant as
<u>Exesipollenites</u> (15% each), and spores make up an even greater
part of the flora (21%). The latter include most of the species
recorded from Drewrys Bluff (except <u>Aequitriradites spinulosus</u>,
<u>Granulatisporites dailyi</u>, <u>Inaperturopollenites pseudoreticulatus</u>,
<u>Klukisporites</u>, and <u>Pilosisporites trichopapillosus</u>), plus many
more, such as <u>Coronatispora valdensis</u> (=<u>Cingulatisporites</u>
<u>caminus</u>), <u>Densoisporites microrugulatus</u>, <u>Lycopodiumsporites</u>
<u>dentimuratus</u>, <u>Microreticulatisporites crassiexinous</u>,
<u>Reticulatasporites dupliexinous</u>, <u>Taurocusporites reduncus</u>, and
<u>T</u>. <u>segmentatus</u>; possibly bryophytic and/or lycopsid spores, such
as <u>Cingulatisporites</u>, <u>Lycopodiacidites</u>, <u>Taurocusporites</u>, and
<u>Triporoletes</u>, including new species not recognized by Brenner
(1963), are especially prominent. In addition, uni- and

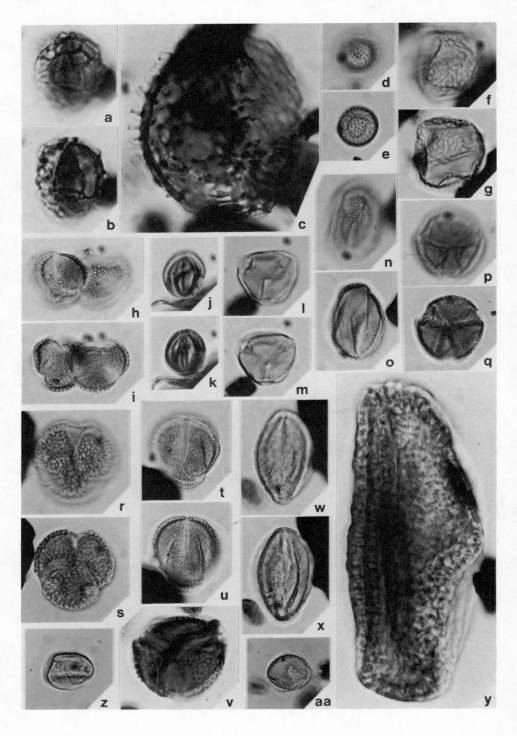

Fig. 4. Pollen and acritarchs from the Dutch Gap (76-27, a-g) and Bodkin Point (76-8, h-y) Pseudofrenelopsis parceramosa localities and the Fredericksburg Frenelopsis ramosissima locality (71-21, z, aa). All light micrographs, X 1000.

a, b: Retimonocolpites peroreticulatus, two focal levels, slide
       76-27-1a, 5.2 X 108.8

c: large, spiny acritarch, 76-27-1a, 5.3 X 110.7

d, e: small, scabrate (echinulate?) acritarch, 76-27-1a,
       7.7 X 111.2

f, g: rugulate acritarch, 76-27-1a, 7.7 X 107.6

h, i: cf. Tricolpites micromunus, two grains, 76-8-1a,
       14.9 X 94.1

j, k: cf. "Tricolporopollenites" distinctus, 76-8-1a,
       18.0 X 105.4

l, m: cf. "Tricolporopollenites" triangulus, 76-8-1a,
       12.2 X 119.2

n, o: aff. Tricolpites nemejci, 76-8-1a, 18.8 X 92.3

p, q: cf. Tricolporoidites bohemicus, 76-8-1a, 11.0 X 116.1

r, s: cf. Tricolpites vulgaris, 76-8-1a, 20.8 X 102.5

t, u: Tricolpites nemejci, 76-8-1a, 20.4 X 93.2

v: Tricolporopollenites sp. A, 76-8-1a, 13.1 X 114.1

w, x: Tricolporoidites sp. C, 76-8-1a, 18.3 X 91.6

y: Stephanocolpites tectorius, 76-8-1a, 13.0 X 90.7

z: smooth, thin-walled acritarch, 71-21-1c, 10.3 X 100.7

aa: small, scabrate (echinulate?) acritarch, 71-21-1c,
     15.1 X 109.1

multicellular fungal remains are diverse and abundant. Acritarchs make up about 1% of the total flora; most are small,
smooth, glassy spheromorphs, but there are also rarer small,
rugulate or scabrate forms (Figs. 4d-g), and one larger type with
forked spines (Fig. 4c).

In terms of original vegetation, the mixed character of the
palynological and megafossil floras at Dutch Gap is most readily
explained as the result of sedimentary influx of pollen, spores,
and (to a predictably lesser extent) macroscopic plant remains
from a rich gymnosperm-fern community into the marginal zone of a
Pseudofrenelopsis parceramosa-dominated, tidally influenced area
where planktonic algae were living in the water. It is tempting
to speculate that Zamites buchianus may have occupied a vegetational zone transitional between nearly pure stands of P.
parceramosa and diverse climax forest communities dominated by
other conifers.

The final Pseudofrenelopsis parceramosa locality, Bodkin
Point, is remarkable in being in the uppermost Potomac Group,
near the contact with the Santonian Magothy Formation, in beds
formerly referred to the Raritan Formation (Berry, 1916). Here
P. parceramosa occurs in a 1 m thick gray silty clay bed, exposed
just above high tide level, which passes laterally (to the south)
into red clays and is overlain by sands. As at the other
localities, the sediment is almost massive, with only a slight
tendency to split along bedding planes, and somewhat mottled,
but it differs in being considerably siltier and lighter in color,
suggesting less reducing environments. The only macroscopic
plant remains are relatively fragmented P. parceramosa stems,
including many single internodes as well as several articulated
specimens (Fig. 3c), a poorly preserved female conifer cone,
several unidentifiable stem fragments, and pieces of fusainized
wood.

The Bodkin Point palynoflora differs markedly from that at
Drewrys Bluff and Dutch Gap, largely as a result of the rise of
angiosperms between the beginning and end of Potomac deposition.
Its qualitative composition permits assignment to the upper part
of Zone III (early Cenomanian?). The angiosperm element includes
relatively undiagnostic forms which range from various horizons
in Zone II through Zone III, such as cf. Tricolpites micromunus
(Figs. 4h,i), cf. T. albiensis, cf. "Tricolporopollenites"
triangulus (Figs. 4l,m), cf. "T." distinctus (Figs. 4j,k), and
various other tricolpates, tricolporoidates, and monosulcates;
other species which first appear near the base of Zone III, such
as cf. Tricolpites vulgaris (Figs. 4r,s), aff. T. nemejci

(Figs. 4n,o), and cf. <u>Tricolporoidites</u> <u>bohemicus</u> (Figs. 4p,q); and a few, such as <u>Tricolpites</u> <u>nemejci</u> (Figs. 4t,u), <u>Tricolporopollenites</u> sp. A (Fig. 4v), <u>Tricolporoidites</u> sp. C. (Figs. 4w,x), and <u>Stephanocolpites</u> <u>tectorius</u> (Fig. 4y) which are restricted to the upper part of Zone III (Doyle and Robbins, 1977). Non-angiospermous elements are also normal for Zone III; e.g., the Albian-Cenomanian spore genus <u>Acritosporites</u> (including <u>Taurocusporites</u> <u>spackmani</u> Brenner: Juhász, 1979), <u>Rugubivesiculites</u> <u>rugosus</u>, which first appears in Subzone II-C (latest Albian?), and other conifers typical of Late Cretaceous floras (<u>Phyllocladidites</u>, <u>Araucariacites</u>, etc.). No Normapolles or other types which enter in Zone IV (middle to late Cenomanian?) have been observed. Since correlations with well-dated sequences in Europe and western North America indicate that upper Zone III falls above the base of the Cenomanian (Doyle and Robbins, 1977), the Bodkin Point occurrence represents a range extension for <u>P. parceramosa</u>, previously known from Berriasian through Albian sediments (Watson, 1977).

Although dominated by angiosperms (36.5%), probable taxodiaceous conifers (23%), and pteridophytes (15.5%), the Bodkin Point palynoflora is noteworthy for the fairly high frequency of <u>Classopollis</u> (13.5%), usually reduced to a few percent by this horizon. <u>Exesipollenites</u> is more subordinate (2%). As at the Zone I localities, fungi are numerous and diverse (more so than in Zone I, perhaps reflecting the continuing Cretaceous radiation of ascomycetes: Pirozynski, 1976), and small, smooth and faintly sculptured acritarchs are present at low frequencies (0.5%). Considering the dominance of <u>P. parceramosa</u> in the megafossil flora, most of the terrestrial palynoflora was probably derived by transport from elsewhere, a conclusion which fits the relatively coarse character of the matrix. It is remarkable that no megafossil remains of the regionally important angiosperm element have been seen at Bodkin Point, suggesting that angiosperms had still not penetrated the sorts of habitat dominated by <u>P. parceramosa</u>.

<u>Frenelopsis</u> <u>ramosissima</u>--This species is known from two Potomac Group localities, Fredericksburg and Baltimore, both of which are assignable to the upper part of Zone I (late Aptian?) on the presence of aff. <u>Retimonocolpites</u> <u>dividuus</u> and the first very rare tricolpates (aff. <u>Tricolpites</u> <u>crassimurus</u>) and the lack of Zone II index species (Doyle and Hickey, 1976; Hickey and Doyle, 1977). Although not very different in age, assemblages from these localities show a markedly different pattern of diversity from those containing <u>Pseudofrenelopsis</u> <u>parceramosa</u> and <u>P. varians</u>. Whereas <u>Pseudofrenelopsis</u> occurs by itself or with at most one other identifiable species, Fredericksburg and

Baltimore were two of the richest known Potomac Group localities,
with a great diversity of well-preserved ferns, equisetalian
stems, cycadophyte fronds (at Fredericksburg), conifer stems and
leaves, and angiosperm leaves (Fontaine, 1889; Berry, 1911;
Wolfe et al., 1975; Doyle and Hickey, 1976; Hickey and Doyle,
1977). Compared with the P. parceramosa localities, Classopollis
makes up an important but not dominant portion of the palynoflora,
fungi are less abundant and diverse, and palynological evidence
for marine influence is inconsistent and ambiguous.

According to Fontaine's (1889) descriptions and examination
of material at the Smithsonian Institution and Harvard, the
Fredericksburg fossils came from a restricted lens of hard,
light to medium gray, micaceous fine sand, packed with organi-
cally preserved plant material lying parallel to the bedding
planes, which occurred in a predominantly conglomeratic sequence.
Although Fontaine's locality is now covered with housing and
vegetation, other exposures in the Fredericksburg area show a
predominance of cross-bedded sands and gravels, including fining-
upward cycles, typical of lower Potomac Group sediments as
described by Glaser (1969). Despite the relatively coarse
character of the matrix, published illustrations (Fontaine, 1889;
Berry, 1911; Watson, 1977) show large, unfragmented specimens of
F. ramosissima with up to five orders of branching and equally
well-preserved remains of other conifers, cycadophytes, ferns,
and angiosperms (58 species in all recognized by Berry, 1911),
as well as comminuted plant fragments, suggesting rapid burial
after a minimum of transport.

Classopollis makes up a higher than average portion of the
Fredericksburg palynoflora (29.5%), but it is associated with a
nearly equal abundance of probable taxodiaceous inaperturates
(30%). Consistent with the diversity of the megafossil flora,
other gymnosperms such as Eucommiidites, Vitreisporites
(Caytoniales), Decussosporites microreticulatus (not its lower
Zone I bisaccate relative), Sciadopitys-like gemmate inaperturates, and psilate monosulcates are present at significant frequencies, and fern spores, especially Cyathidites and
Schizaeaceae, make up 19% of the palynoflora. (Any numbers of
these could, of course, have been transported from farther
upstream, considering the coarseness of the matrix). In further
contrast to the Pseudofrenelopsis localities, Exesipollenites
is present only as isolated grains, and fungi include only
relatively inconspicuous unicellular forms and masses rather
than chains. Very sporadic small, smooth to scabrate spheres
(Figs. 4z,aa) occur which may represent plankton, but they
contrast with the spherical acritarchs seen at the P. parceramosa
localities in their thinner and apparently less rigid, more
readily crushed walls and less glassy appearance. They are

hence difficult to distinguish from paler unicellular fungal
spores in the same sample, raising the possibility that they are
of fungal rather than planktonic origin; the scabrate forms are
suggestive of certain myxomycete spores illustrated by Graham
(1971).*

The Baltimore material of Frenelopsis ramosissima is from
relatively light gray, silty clays, overlying lower Potomac Group
sands, which were formerly exposed near the base of Federal Hill
(Fontaine, 1889), now the site of a park. Although assigned to
the Patapsco Formation by Berry (1911), palynology and regional
geologic relations indicate that these beds correlate with the
Arundel Clay (upper Zone I, late Aptian?), although they contain
more silt and show better bedding than is typical for the Arundel.
The megafossil flora is rich in ferns, conifers, and angiosperms,
although less so than that at Fredericksburg (28 species recognized by Berry, 1911), from which it also differs in the apparent
lack of cycadophytes and the species composition of the angiosperm
element (cf. Doyle and Hickey, 1976). Some of the angiosperms
have been interpreted as possible semiaquatic herbs (Doyle and
Hickey, 1976; Hickey and Doyle, 1977), particularly Plantaginopsis
marylandica, with elongate leaves arranged in a rosette, and
Vitiphyllum multifidum, with lobate leaves with unbraced sinuses,
which would be consistent with the paludal environment for the
Arundel proposed by Glaser (1969) and others.

The Baltimore palynoflora shows the same lack of monodominance noted at Fredericksburg. Classopollis is the most
common genus (20%), but it is little more abundant than the Zone
I average (16%), and it is associated with relatively abundant
inaperturate conifers (16%), bisaccates (11.5%), Cyathidites
(10%), Eucommiidites (9.5%), Exesipollenites (7.5%), and
schizaeaceous spores (5.5%). Much of this diversity could be the
result of sedimentary mixing from various sources, but the
megafossil flora indicates that much of it probably reflects
diverse nearby vegetation. As at Fredericksburg, fungal remains
are relatively uncommon and consist mostly of unicellular spores.
However, the probable planktonic element is more heterogeneous,
with both occasional small, smooth, spherical acritarchs of the
sort found at the P. parceramosa localities (1%) and Schizosporis
reticulatus, a large, possibly colonial form recently compared by
Pierce (1977) with a modern freshwater alga. Considering the
uncertainties on systematic affinities and ecological implica-
tions of simple acritarchs, it is impossible to say whether this
association is the result of mixing of freshwater and brackish-
marine elements or evidence for the freshwater nature of some of
the smooth acritarchs.*

*
See note added in proof, p. 196.

Table 1. Frequencies of major spore, pollen, and plankton groups
in Potomac Group and Glen Rose samples studied. Spore and pollen
percentages are of terrestrial sum (200 specimens); plankton percentages, of terrestrial plus planktonic sum. x = observed in
sample but not encountered in count.

	76-21 (Drewrys Bluff)	76-27 (Dutch Gap)	76-8 (Bodkin Point)	71-21 (Fredericksburg)	71-6 (Baltimore)	79-26 (Glen Rose)	OG 1323 (Oak Grove)
Cyathidites	0.5	2.5	1	7	10	x	x
Gleicheniidites	--	0.5	2	x	1	x	0.5
Striate Schizaeaceae	0.5	5.5	2.5	4.5	5.5	x	0.5
Other spores	3	12.5	10	7.5	5.5	3	2
Classopollis	78	21	13.5	29.5	20	78	60.5
Exesipollenites tumulus	10.5	16.5	2	x	7.5	10	15.5
Inap. dubius (Taxodiaceae?)	4.5	15	23	30	16	6	15
Bisaccate conifers	1.5	15	7	4	11.5	x	3.5
aff. Decussosporites	x	4.5	--	--	--	--	1
Vitreisporites	x	0.5	--	4	0.5	x	0.5
Eucommiidites	x	1.5	1	2	9.5	0.5	0.5
Smooth monosulcates	1	2.5	1.5	4	2.5	x	0.5
Other gymnosperms	x	2.5	x	6.5	7.5	0.5	x
Angiospermous monosulcates	0.5	x	0.5	0.5	2.5	2	x
Tricolpates, tricolporates	--	--	36	0.5	x	x	--
Acritarchs	0.5	1	0.5	0.5	1	6.5	2
Dinoflagellates	--	--	--	--	--	14.6	--
Foraminifera	--	--	--	--	--	1.9	--

DISCUSSION

Because of the inferred complexity and heterogeneity of Potomac Group depositional environments and uncertainties in evaluating the significance of low frequencies of nondescript acritarchs, it is impossible to draw simple, unambiguous conclusions on levels of salinity tolerated by Pseudofrenelopsis parceramosa and Frenelopsis ramosissima. However, when considered in conjunction, sedimentological features, megafossil associates, palynology, and comparisons with palynological and megafossil assemblages elsewhere indicate marked differences in the ecology of the two species and have clear relevance to hypotheses on the ecology of Cheirolepidiaceae in general.

In the case of P. parceramosa, various lines of evidence-- low diversity or absence of associated plant megafossils, occurrence in silty gray clays often showing evidence of bioturbation, and association with a low-diversity palynofacies characterized by dominance of Classopollis and Exesipollenites, common fungi, and low frequencies of morphologically simple acritarchs--are highly consistent with the concept of this species as a dominant of coastal, tidally influenced habitats and a facultative (though not necessarily obligate) halophyte. Morphologically, it is more analogous to certain modern salt marsh shrubs (e.g., Salicornia) than to modern mangroves, which are less xeromorphic in their vegetative morphology. The occurrences of P. parceramosa in the Potomac Group and P. varians in the Glen Rose Formation of Texas both show that high frequencies of Classopollis in marginal marine sediments do not necessarily reflect contributions from an upland source, whatever the validity of the Neves Effect (Chaloner and Muir, 1968) in other cases or the role of other Cheirolepidiaceae in upland vegetation. However, the far more abundant, diverse, and unambiguously marine planktonic element in the Glen Rose could mean that Potomac Group P. parceramosa was adapted to less saline (brackish?) conditions than P. varians. Furthermore, the sedimentological differences seen at the three Potomac localities suggest that P. parceramosa may have occupied a certain range of environments, at Dutch Gap possibly adjacent to richer conifer-fern-cycadophyte communities. It is also possible to imagine that P. parceramosa was growing at the margins of marshes in nonsaline conditions, while the acritarchs associated with its remains after burial were restricted to denser salt water at the bottom, brought in by tidal action. Finally, it would be premature to exclude the possibility that P. parceramosa extended into special nonsaline habitats where its xeromorphic features were adaptive (cf. Alvin et al., 1978, and below).

In contrast, none of the present data would suggest that
F. ramosissima dominated tidally influenced, coastal environments.
This species consistently occurs as a member of highly diverse,
well-preserved megafossil floras which include typically non-
halophytic groups such as ferns and Equisetales as well as
various non-xeromorphic gymnosperms and angiosperms. The
lithofacies at its two localities are rather dissimilar--
relatively fine at Baltimore, but coarse and fluvial at
Fredericksburg--but also different from facies containing P.
parceramosa. Especially at Fredericksburg, where both branch
systems of F. ramosissima and other large, often delicate plant
remains are well preserved in a relatively coarse matrix, it is
difficult to imagine F. ramosissima as the one local coastal
marsh element and the other plants as transported in. The
associated palynoflora also contrasts with the low-diversity,
presumptively marine-influenced Classopollis-Exesipollenites-
fungal-acritarch facies seen at the P. parceramosa localities:
Classopollis is abundant but not dominant, Exesipollenites is
barely present at one of the localities (Fredericksburg), fungi
are subordinate, and possible planktonic evidence of marine
influence is problematical and inconsistent at the two localities.

As an alternative, non-halophytic explanation for the
peculiar morphology of Frenelopsis ramosissima, we would propose
an analogy with certain xeromorphic plants which occur in other-
wise mesophytic vegetation of the modern humid tropics and subtropics. In these regions, xeromorphy is characteristic of
plants that grow on rock outcrops (such as cacti: Janzen, 1975)
and sandy, porous soils; examples of the latter are the Heath
Forest of Borneo, the Wallaba Forest of Guyana (dominated by the
legume Eperua falcata), and the padang vegetation of Malaysia
(Richards, 1952). These communities exhibit a higher species
diversity than most mangrove swamps (cf. Richards, 1952), a fact
consistent with such an ecological interpretation for F.
ramosissima. Cases such as these, along with the more familiar
one of xeromorphy in plants of temperate bogs, have led to the
concept of xeromorphy as characteristic of plants which must
cope with a deficiency in one or more essential elements, and not
just drought or osmotic stress (Walter, 1979; Givnish, 1979). In
fact, Givnish predicts that arid climates, effectively dry
habitats, mineral-poor soils, and soils with low oxygen concentrations should all select for small, thick leaves, since in
each case transpiration costs are high relative to photosynthetic
profits. This means that xeromorphy relative to other members of
a flora cannot be used in isolation as an indication that a
particular species tolerated elevated salinities.

In view of the recent suggestion that early angiosperms were coastal halophytes (Retallack and Dilcher, 1979), it is worth noting that angiosperm leaves occur at both localities with <u>F</u>. <u>ramosissima</u>, but they have not been found despite intensive search in any of the beds containing <u>P</u>. <u>parceramosa</u>, not even at Bodkin Point, which represents a time when angiosperms had already become dominant in many other facies (cf. Hickey and Doyle, 1977).

Our study suggests several avenues of future research which could further elucidate the ecology of the Cheirolepidiaceae and improve our understanding of terrestrial paleoenvironments in general. First, more detailed studies are needed on the abundance and diversity of acritarchs in modern and ancient fluvial-deltaic sediments, particularly along independently determined salinity gradients. Some previous investigations, such as Muller's (1959) pioneering study on the modern Orinoco delta, have reported no marine microfossils even at some distance from the shoreline, while in other cases (e.g., the Glen Rose) dino-flagellates, acritarchs, and foraminifera are abundant in supposed intertidal deposits. The rarity and simple morphology of the acritarchs in many of our samples suggest that micro-plankton may often have been overlooked in nearshore marine and brackish environments. Second, the different assemblages of fungal remains in our samples suggest that fungi may prove to be sensitive paleoenvironmental indicators (cf. Pirozynski, 1976). In modern salt marshes and mangrove swamps, those species which attack plant parts below tidal level are almost always restricted to marine waters, while those which attack the emergent portions are terrestrial in distribution (Kohlmeyer and Kohlmeyer, 1979). Third, further research is needed to document the distribution of xeromorphy in the humid tropics and subtropics, its different forms (e.g., stem succulence vs. sclerophyllous leaves), and its relations to ecological factors. Lastly, more studies are desirable which analyze megafossil abundance and preservation along marine to continental gradients in correlative sediments. Such research should help determine which members of the Cheirolepidiaceae were restricted to coastal environments and which had more inland distributions.

In conclusion, we believe that the very different environ-mental conditions inferred from palynological, megafossil, and sedimentological associations of <u>Frenelopsis</u> <u>ramosissima</u>, <u>Pseudofrenelopsis</u> <u>parceramosa</u>, and <u>P</u>. <u>varians</u> support the concept of Batten (1974, 1976) and Alvin, Spicer, and Watson (1978) that the Cheirolepidiaceae as a whole were extremely varied in their ecological tolerances, occupying a wide range of coastal to inland habitats in the Mesozoic tropical and subtropical zones. Until it is proven on a case by case basis that certain members

were strictly coastal plants or obligate halophytes, their
presence cannot be used by itself as evidence of coastal vegetation or marine influence.

NOTE ADDED IN PROOF

In writing this paper, we were unable to·determine whether
low-diversity plankton assemblages consisting largely of smooth,
spherical acritarchs indicate freshwater conditions or low
levels of marine influence, and hence whether or not the
occurrence of such assemblages at the Frenelopsis ramosissima
localities is consistent with our concept that F. ramosissima
was not a coastal halophyte. However, new evidence strengthens
our interpretation by showing that at least some such acritarch
assemblages are of freshwater origin. Since submitting this
paper, we have observed occasional smooth, spherical acritarchs
similar to those in Figs. 2g,h and 4z, plus rarer glassy
spheres with fine spines, in a palynological preparation from
andesitic sands of late Miocene age (formerly considered
Pliocene) near Verdi, Nevada (Axelrod, D. I., 1958, The Pliocene
Verdi flora of western Nevada, Univ. Calif. Publ. Geol. Sci.
34: 91-160; Axelrod, D. I., 1980, Contributions to the Neogene
paleobotany of central California, Univ. Calif. Publ. Geol. Sci.
121: 1-212). Since these sediments were deposited east of the
crest of the Sierra Nevada, at elevations estimated by Axelrod
from the included megafossil flora and other evidence as about
2000', and since the flora indicates conditions too humid to
support saline lakes, it is clear that these acritarchs represent freshwater forms. Hence, although it is still possible that
the more diverse acritarch assemblages at the Drewrys Bluff and
Dutch Gap Pseudofrenelopsis parceramosa localities constitute
evidence for some level of marine influence, the types of
acritarchs at the Baltimore and Fredericksburg F. ramosissima
localities, and even the Bodkin Point P. parceramosa locality,
are wholly consistent with, if not indicative of, freshwater
conditions.

ACKNOWLEDGEMENTS

We wish to thank Leo Hickey, Jeffrey Mount, Robert Pearcy,
and George Rogers for invaluable discussions of sedimentological
and ecological problems, and Gilbert Brenner and Carol Hotton
for help in the field. Field work was supported by the Museum
of Paleontology, University of Michigan.

REFERENCES

Alvin, K. L., and Pais, J. J. C. 1978. A Frenelopsis with
    opposite decussate leaves from the Lower Cretaceous of
    Portugal. Palaeontology 21: 873-879.

Alvin, K. L., Spicer, R. A., and Watson, J. 1978. A
    Classopollis-containing male cone associated with
    Pseudofrenelopsis. Palaeontology 21: 847-856.

Anderson, J. L. 1948. Cretaceous and Tertiary subsurface
    geology. Maryland Dept. Geol. Mines Water Res. Bull. 2:
    1-113.

Barnard, P. D. W. 1968. A new species of Masculostrobus
    Seward producing Classopollis pollen from the Jurassic of
    Iran. J. Linn. Soc. (Bot.) 61: 167-176.

-------. 1973. Mesozoic floras. In Organisms and continents
    through time, N. F. Hughes, ed., Palaeont. Assoc. London
    Spec. Papers Palaeont. 12: 175-187.

Basan, P. B., ed. 1978. Trace fossil concepts. SEPM short
    course no. 5, Soc. Econ. Paleont. Mineral., Oklahoma
    City, 201 pp.

Batten, D. J. 1974. Wealden palaeoecology from the distribution
    of plant fossils. Proc. Geol. Assoc. 85: 433-458.

-------. 1976. Correspondence: Wealden of the Weald--a new
    model. Proc. Geol. Assoc. 87: 431-433.

Berry, E. W. 1911. Pteridophyta--Dicotyledonae. In Lower
    Cretaceous, Maryland Geol. Surv., Johns Hopkins Press,
    Baltimore, 214-508.

-------. 1916. Fossil plants. In Upper Cretaceous, Maryland
    Geol. Surv., Johns Hopkins Press, Baltimore, 757-901.

Brenner, G. J. 1963. The spores and pollen of the Potomac
    Group of Maryland. Maryland Dept. Geol. Mines Water Res.
    Bull. 27: 1-215.

-------. 1976. Middle Cretaceous floral provinces and early
    migrations of angiosperms. In Origin and early evolution
    of angiosperms, C. B. Beck, ed., Columbia Univ. Press,
    New York, 23-47.

Chaloner, W. G. 1958. The Carboniferous upland flora. Geol. Mag. 95: 261-262.

Chaloner, W. G., and Muir, M. 1968. Spores and floras. In Coal and coal-bearing strata, D. G. Murchison and T. S. Westoll, eds., Oliver and Boyd, Edinburgh, 127-146.

Clapham, W. B. 1970. Nature and paleogeography of Middle Permian floras of Oklahoma as inferred from their pollen record. J. Geol. 78: 153-171.

Clark, W. B., Bibbins, A. B., and Berry, E. W. 1911. The Lower Cretaceous deposits of Maryland. In Lower Cretaceous, Maryland Geol. Surv., Johns Hopkins Press, Baltimore, 23-98.

Daghlian, C. P., and Person, C. P. 1977. The cuticular anatomy of Frenelopsis varians from the Lower Cretaceous of central Texas. Am. J. Bot. 64: 564-569.

Delteil, J. R., Le Fournier, J., and Micholet, J. 1975. Schéma d'évolution sédimentaire d'une marge continentale stable: exemple-type du Golfe du Guinée de l'Angola au Cameroun. 9e Congr. Internat. Sédimentol., Nice 1975, Thème 4, 1: 91-96.

Doludenko, M. P., and Reymanówna, M. 1978. Frenelopsis harrisii sp. nov. from the Cretaceous of Tajikistan, USSR. Acta Palaeobot. 19: 3-12.

Doyle, J. A. 1969. Cretaceous angiosperm pollen of the Atlantic Coastal Plain and its evolutionary significance. J. Arnold Arbor. 50: 1-35.

Doyle, J. A. 1977. Spores and pollen: the Potomac Group (Cretaceous) angiosperm sequence. In Concepts and methods of biostratigraphy, E. G. Kauffman and J. E. Hazel, eds., Dowden, Hutchinson and Ross, Stroudsberg, PA., 339-363.

Doyle, J. A., Biens, P., Doerenkamp, A., Jardiné, S. 1977. Angiosperm pollen from the pre-Albian Lower Cretaceous of Equatorial Africa. Bull. Cent. Rech. Explor.-Prod. Elf-Aquitaine 1: 451-473.

Doyle, J. A. and Hickey, L. J. 1976. Pollen and leaves from the mid-Cretaceous Potomac Group and their bearing on early angiosperm evolution. In Origin and early evolution of angiosperms, C. B. Beck, ed., Columbia Univ. Press, New York, 139-206.

Doyle, J. A., and Robbins, E. I. 1977. Angiosperm pollen
    zonation of the continental Cretaceous of the Atlantic
    Coastal Plain and its application to deep wells in the
    Salisbury Embayment. Palynology 1: 43-78.

Drake, A. A., Nelson, A. E., Force, L. M., Froelich, A. J., and
    Lyttle, P. T. 1979. Preliminary geologic map of Fairfax
    County, Virginia. U.S. Geol. Surv. Open File Rept. 79-398,
    2 sheets.

Filatoff, J. 1975. Jurassic palynology of the Perth Basin,
    Western Australia. Palaeontographica Abt. B 154: 1-113.

Fontaine, W. M. 1889. The Potomac of Younger Mesozoic flora.
    U.S. Geol. Surv. Monogr. 15: 1-375.

Givnish, T. 1979. On the adaptive significance of leaf form.
    In Topics in plant population biology, O. T. Solbrig, S.
    Jain, G. B. Johnson, and P. H. Raven, eds., 375-407.

Glaser, J. D. Petrology and origin of Potomac and Magothy
    (Cretaceous) sediments, middle Atlantic Coastal Plain.
    Maryland Geol. Surv. Rept. Investig. 11: 1-102.

Graham, A. 1971. The role of Myxomyceta spores in palynology
    (with a brief note on the morphology of certain algal
    zygospores). Rev. Palaeobot. Palynol. 11: 89-99.

Harris, T. M. 1965. Dispersed cuticles. Palaeobotanist 14:
    102-105.

-------. 1974. Williamsoniella lignieri, its pollen and the
    compression of spherical pollen grains. Palaeontology 17:
    125-148.

Hickey, L. J., and Doyle, J. A. 1977. Early Cretaceous fossil
    evidence for angiosperm evolution. Bot. Rev. 43: 2-104.

Hluštík, A., and Konzalová, M. 1976. Polliniferous cones of
    Frenelopsis alata (K. Feistm.) Knobloch from the Cenomanian
    of Czechoslovakia. Vešt. Ústřed. Ust. Geol. 51: 37-45.

Hughes, N. F. 1973. Mesozoic and Tertiary distributions, and
    problems of land-plant evolution. In Organisms and con-
    tinents through time, N. F. Hughes, eds., Palaeont. Assoc.
    London Spec. Papers Palaeont. 12: 188-198.

-------. 1976. Palaeobiology of angiosperm origins. Cambridge
    Univ. Press, 242 pp.

Hughes, N. F., Drewry, G. E., and Laing, J. F. 1979. Barremian
    earliest angiosperm pollen. Palaeontology 22: 513-535.

Janzen, D. 1975. Ecology of plants in the tropics. E. Arnold,
    London, 66 pp.

Jardiné, S., Doerenkamp, A., and Biens, P. 1974. Dicheiropollis
    etruscus, un pollen caractéristique du Crétacé inférieur
    afro-sudaméricain. Conséquences pour l'évaluation des
    unités climatiques et implications dans la dérive des
    continents. Sci. Géol. Bull. Strasbourg 27: 87-100.

Juhász, M. 1979. Investigation of some spore genera from the
    Lower and Middle Cretaceous in Transdanubia. Acta Biol.
    Szeged 25: 49-64.

Jung, W. 1974. Die Konifere Brachyphyllum nepos Saporta aus
    den Solnhofener Plattenkalk (unteres Untertithon), ein
    Halophyt. Mitt. Bayer. Staatsamml., Paläont. Hist. Geol.
    14: 49-58.

Kohlmeyer, J., and Kohlmeyer, E. 1979. Marine mycology: the
    higher fungi. Academic Press, New York, 690 pp.

Müller, H. 1966. Palynological investigations of Cretaceous
    sediments in northeastern Brazil. Proc. 2nd West Afr.
    Micropaleont. Colloq., Ibadan 1965, J. E. van Hinte, ed.,
    Brill, Leiden, 123-136.

Muller, J. 1959. Palynology of Recent Orinoco delta and shelf
    sediments. Micropaleontology 5: 1-32.

Nagle, J. S. 1968. Glen Rose cycles and facies, Paluxy River
    Valley, Somervell County, Texas. Texas Bur. Econ. Geol.
    Circ. 68-1, 25 pp.

Owens, J. P. 1969. Coastal Plain rocks. In The geology of
    Harford County, Maryland, Maryland Geol. Surv.,
    Baltimore, 77-103.

Perkins, B. F., and Langston, W. 1979. Lower Cretaceous
    shallow marine environments in the Glen Rose Formation:
    dinosaur tracks and plants. Am. Assoc. Strat. Palynol.
    12th Ann. Mtg. Field Trip Guide, 55 pp.

Pettitt, J. M., and Chaloner, W. G. 1964. The ultrastructure
    of the Mesozoic pollen Classopollis. Pollen Spores 6:
    611-620.

Pierce, S. T.   1977.   A modern analog of <u>Schizosporis</u> <u>reticulatus</u>.
    Palynology 1: 139-142.

Pirozynski, K. A.   1976.   Fossil fungi.   Ann. Rev. Phytopathol.
    14: 237-246.

Pirozynski, K. A., and Weresub, L. K.   1979.   A biogeographic
    view of the history of Ascomycetes and the development of
    their pleomorphism.   <u>In</u> The whole fungus, B. Kendrick, ed.,
    Proc. 2nd Internat. Mycol. Conf., Kananaskis, v. 1,
    93-123.

Pocock, S. A. J., and Jansonius, J.   1961.   The pollen genus
    <u>Classopollis</u> Pflug, 1953.   Micropaleontology 7: 439-449.

Pons, D.   1979.   Les organes reproducteurs de <u>Frenelopsis</u> <u>alata</u>
    (K. Feistm) Knobloch, Cheirolepidiaceae du Cénomanien de
    l'Anjou, France.   C. R. 104e Congr. Nat. Soc. Sav.,
    Bordeaux 1979, Sci., fasc. 1, 209-231.

Pons, D., Lauverjat, J., and Broutin, J.   1980.   Paléoclimatologie
    comparée de deux gisements du Crétacé supérieur d'Europe
    occidentale.   Mém. Soc. Géol. France, N.S. 139: 151-157.

Reinhardt, J., Christopher, R. A., and Owens, J. P.   1980.   Lower
    Cretaceous stratigraphy of the core.   <u>In</u> Geology of the Oak
    Grove core.   Virginia Div. Min. Res. Publ. 20: 31-52.

Retallack, G., and Dilcher, D. L.   1979.   The coastal theory of
    flowering plant origins, dispersal and rise to dominance.
    Bot. Soc. Am. Misc. Ser. Publ. 157: 36 (abstr.).

Reyment, R. A., and Tait, E. A.   1972.   Biostratigraphical dating
    of the early history of the South Atlantic Ocean.   Philos.
    Trans. Royal Soc. London Ser. B (Biol. Sci.) 264: 55-95.

Reyre, D., Belmonte, Y., Derumaux, F., and Wenger, R.   1966.
    Evolution géologique du Bassin Gabonais.   <u>In</u> Sedimentary
    basins of the African coasts, Part 1, Atlantic Coast, D.
    Reyre, ed., Assoc. Afr. Geol. Surveys, Paris, 171-191.

Richards, P. W.   1952.   The tropical rain forest.   Cambridge
    Univ. Press, 450 pp.

Upchurch, G. R.   1978.   A preliminary report on the cuticular
    structure of the oldest known structurally preserved
    angiosperm leaves.   Bot. Soc. Am. Misc. Ser. Publ. 156:
    78 (abstr.).

Vakhrameev, V. A.   1970.   Zakonomernosti rasprostraneniya i
    paleoekologiya mezozoyskikh khvoynykh Cheirolepidiaceae.
    Paleont. Zh. 1970: 19-34.

-------.   1978.   Klimaty Severnogo polushariya v melovom periode
    i dannye paleobotaniki.   Paleont. Zh. 1978: 3-17.

Wall, D.   1965.   Microplankton, pollen, and spores from the
    Lower Jurassic in Britain.   Micropaleontology 11: 151-190.

Walter, H.   1979.   Vegetation of the earth and ecological
    systems of the geo-biosphere.   Springer-Verlag, New York,
    274 pp.

Watson, J.   1977.   Some Lower Cretaceous conifers of the
    Cheirolepidiaceae from the U.S.A. and England.
    Palaeontology 20: 715-749.

Wolfe, J. A., Doyle, J. A., and Page, V. M.   1975.   The bases
    of angiosperm phylogeny: paleobotany.   Ann. Missouri
    Bot. Gard. 62: 801-824.

THE STRUCTURAL AND PHYTOGEOGRAPHIC AFFINITIES OF SOME SILICIFIED

WOOD FROM THE MID-TERTIARY OF WEST-CENTRAL MISSISSIPPI

Will H. Blackwell, David M. Brandenburg,
and George H. Dukes*

Miami University
Oxford, Ohio 45056

ABSTRACT

Pieces of fossil "wood" were collected with special permission from the "petrified forest" located near the town of Flora, Mississippi (Madison County), during the summers of 1977 and 1979. This material was subsequently studied in laboratories at Miami University. The "forest" has long been known in the literature, and was finally designated a National Natural Landmark in 1966. Although the site has received both popular and scientific attention over the years, little of scientific substance has been published concerning it. Initially there was confusion over the nature of the formation (Forest Hill Sand); however, it is considered to be Lower Oligocene or on the Eocene/Oligocene boundary, and of shallow water (nonmarine) origin. The fossilized plant material (primarily tree trunks) is silicified, apparently a consequence of gradual mineral infiltration following rapid burial in soft sand. Chief replacement minerals are low temperature silicates: chalcedony and opal. Although much fossil wood was removed prior to conservation, trunk segments continue to surface from the eroding sands. Chips are rather easily obtained, and are well suited to thin-section techniques. Cellular preservation varies but is quite often good. Much confusion has existed as to the types of wood present; however, repeated sampling has demonstrated the presence of only two: one gymnospermous, the

---

*
 Present Address: Mississippi State Department of Education,
 Box 771, Jackson, MS 39205.

Fig. 1: Forest Hill Formation, "Badlands" topography; note petrified logs in foreground.

Fig. 2: Petrified log emerging from side of formation.

Fig. 3: Weathered surface of large petrified log.

Fig. 4: Several petrified logs lying on surface, giving illusion of "log jam".

Fig. 5: Large petrified log segment, "Caveman's Bench," often featured on postcards or brochures at Mississippi Petrified Forest.

other angiospermous. The gymnosperm petrifaction is harder and
shows less internal decay. Scattered axial parenchyma, absence
of resin ducts, tracheid shape, and cross-field pitting indicate
the wood to be cupressoid. The nearest extant counterpart occurs
in Northwest Africa. On the other hand, the angiosperm material
has presented greater difficulty in determination. However, the
extremely heterogeneous rays and other aspects of the wood bear
a resemblance to certain of the arborescent Euphorbiaceae. The
phytogeographic relationships of both woods are apparently
tropical. There is little to substantiate the "driftwood from
the north" hypothesis previously offered for the origin of these
petrified logs. More likely, they were part of a swamp forest
existing in close proximity to the site of fossilization. Both
woods are described in this paper.

INTRODUCTION

     The existence of a petrified forest in Mississippi has been
known for some time (Brown, 1913), and accounts have appeared in
such popular writings as National Geographic (Hildebrand, 1937),
the American Guide Series (Federal Writers' Project, 1938), and
the National Encyclopedia (Suzzallo, 1932). For many years little
was done to conserve this valuable assemblage of fossil logs,
and many were removed from the site piece by piece. However,
through the efforts of the Schabilion family (who purchased the
petrified forest), the location became registered as a National
Natural Landmark in 1966, and collection or removal of any fossil
material has been illegal since, without special permission.
Fortunately, a fair number of logs (some of considerable size)
still remain (Figs. 3, 4 and 5) and more continue to emerge
intermittently from the deposit in which they are buried (Fig. 2).
The Schabilion family has been instrumental not only in preserving
this interesting material, but also in displaying it to the pub-
lic. They have provided several informative brochures describing
it, as well as a write-up in Rock & Gem magazine (Schabilion,
1973).

     The Mississippi Petrified Forest lies approximately three
miles to the southwest of the town of Flora, Mississippi (Madison
County), in connection with a geologic formation known as the
Forest Hill Sand (Wilmarth, 1938). The Forest Hill Formation is
considered Lower Oligocene in age (36-38 million years B.P.), or
has been interpreted by some to occur on the Eocene/Oligocene
boundary (Berry, 1924). It is somewhat differentiated lithologi-
cally from the Jackson (Upper Eocene) Formation below and the
Vicksburg Group (Lower Oligocene) with which it is sometimes
correlated (Wilmarth, 1938) by the fact that the Forest Hill is a
shallow water (nonmarine) deposit. Debates over the geologic age
of the formation are admirably reviewed by Ainsworth (1967).

The reddish to yellow-white or pale brown sands of which it is
composed lie in a general amphitheater (of undetermined horizontal
extent) as much as 75 feet deep in places. As indicated, a
number of logs lie exposed on the surface; however, some continue
to surface from time to time from the rapidly erodable sands
("badlands topography", Figs. 1 and 2).

    Botanically, although little has been published, a reasonable
amount of scientific attention has been given to the fossil wood
at Flora. Rapp (1954) concluded that Shaw (1918) must have sent
thin-sections of wood from Flora to F. H. Knowlton who identified
one as a conifer bearing a resemblance to Sequoia and the other
as a dicotyledon showing similarity to Quercus. It is clear at
least that two types of wood were recognized at Flora fairly
early in the literature. Rapp, who studied the woods at Flora
as part of his master's thesis, also recognized two types,
assigning the coniferous wood to Cupressinoxylon (although cross-
field pitting was "not observed") and the angiosperm wood to
Floroxylon, a [form] genus of uncertain family affinity. The
woods at Flora were also the subject of a master's thesis by
George Dukes (1959) who recognized a range of conifer and dicot
wood types. Dukes (1961a) followed this work up as a part of a
doctoral dissertation, but declined there to pronounce on the
exact identifications of the Flora woods. An abstract of Dukes'
doctoral work was published (1961b) in which he described the
Flora forest woods as being "of 1 type, possibly 2". The only
published photograph of microscopic sections of the Flora woods
is apparently taken from Dukes (1959) and published in Priddy's
(1960) geological survey of Madison County. In this photograph,
the conifer is identified as Abies and the angiosperm as Acer.
The general presumption has been that the fossil logs at Flora
were of northern affinity (Schabilion, 1973), having been transported to Mississippi as driftwood of distant origin. The
weathered appearance of some of the fossil logs (Fig. 3), a
supposed arrangement in a "log jam" (Fig. 4), and identifications
such as sequoia, fir, and maple, quite understandably contributed
to such a concept. However, in one of the brochures disseminated
by the Schabilions nicely depicting the fossil forest at Flora,
there is mention of the possibility that one of the logs might
be a "spurge". This resulted from an opinion offered by
Virginia Page of Stanford University, based on sectioning of
wood samples sent to her by a member of the Schabilion family;
however, nothing of this nature was published other than in one
or more of the publicity brochures. The petrified forest at
Flora has received essentially no study since the early 1960's
until our investigations began in 1977.

    With regard to the mechanism of fossilization, the assump-
tion (probably correct) has been that the logs underwent rapid

burial in the soft (perhaps marshy) sands, which prevented total
deterioration. Subsequently, they gradually became silicified as
a result of replacement of the organic material by various
minerals such as "minute quartz crystals" (Schabilion, 1973, and
various handout brochures). Dukes (1961a) alluded to the silici-
fication as involving impregnation and replacement, and on the
basis of some experimentation (using hydrofluoric acid digestion)
found no evidence for remains of cell wall material such as
cellulose or lignin. The fossil logs at Flora do appear to be
total replacements, or what might be termed simulation petrifactions. However, very little has been published concerning the
actual replacement minerals beyond the general statement that
they are silicates. Thus, in addition to identification of the
woods, the minerology was construed to be an integral part of our
investigation.

MATERIALS AND METHODS

With special permission of the Schabilion family, to whom
we are deeply indebted for their cooperation, modest though wide-
ranging collections of fossil wood were made in the summers of
1977 and 1979. This material was subsequently transported to
Miami University, Oxford, Ohio, where work was done in geological
laboratories. We are indebted to the Department of Geology for
making their facilities available to us.

Chips of fossil wood (at least several centimeters in
diameter each) were sectioned with a diamond (rock) saw so as
to fit on petrographic slides. For each type of wood, three types
of sections (cross, radial-longitudinal, and tangential-longitu-
dinal) were obtained. After sectioning, the slugs of fossil wood
were polished with a silicon carbide grit series on a lapidary
wheel, fine-polished on a glass plate (using 1000 grit), rinsed,
dried, and then mounted with Hillquist epoxy on the petrographic
slides. Upon hardening of the resin (overnight, under pressure)
the cemented wood slugs were cut again with a thin-section saw
(as thin as possible) and polished to the desired transparency
(typically less than 50 micrometers) on a grinding wheel. After
drying these thin-ground sections, coverslips were mounted on
them (with pressure) using the same (Hillquist) medium and permitted to set, usually under heat (80°C for one hour).

Peel techniques proved unsatisfactory for this type of work
because of the need to observe fine cell detail in some depth
of focus.

After appropriate coding, the thin sections were examined
for botanical determination by ordinary light microscopy. Photo-
micrography was accomplished with an AO-Microstar microscope in

combination with an AO camera back or a Polaroid Land microscope
camera. However, polarizing microscopy (Leitz polarizing microscope) proved essential for mineral identification.

RESULTS AND DISCUSSION

Cellular preservation (by mineral replacement) of the fossil
woods at Flora is generally quite good, with reproduction faith-
ful in detail. In our preparations it was usually rather easy to
avoid decayed areas, apparently caused by a Basidiomycete
(identified during our microscopical investigations by Dr. Martha
Powell) prior to siliceous replacement (i.e., 36-38 million years
ago). Somewhat in correspondence with previous investigations,
two types of silicified wood, and two only, were found to be
present in the Forest Hill Formation. One of these (generally
the better preserved) is definitely coniferous, and the other
(apparently more abundant) dicotyledonous. Although woods more
or less similar to these have been properly described by various
workers from a number of different localities in the southeastern
United States (e.g., Knowlton, 1891; Penhallow, 1907; Berry, 1922),
the Flora logs per se have never been legitimately (i.e., effectively and therefore validly) described in a nomenclatural sense
(cf. Stafleu et al., 1978), and the angiospermous material has
not been adequately designated in terms of botanical relationships. Additionally, there is the confusion of the various
misidentifications of the material in the literature. Thus
below, hopefully, we will succeed in "legitimizing" the correct
designation for each wood type. Authentic (holotypic) material
for both woods described is deposited in the paleobotanical
collections of the Herbarium of Miami University (MU). No fossil
plant material other than petrified wood was found to occur at
Flora.

Botanical Determinations and Descriptions

CUPRESSACEAE:

CUPRESSINOXYLON florense Rapp ex Blackwell in Blackwell et al.,
        sp. nov. (see Goeppert, 1850, for generic description of
        Cupressinoxylon, and Berry, 1922, for English translation;
        see Rapp, 1954, for proposal of species name Cupressinoxylon
        florense, as "Cupressoxylon" florense)

Transverse Section (Fig. 6): Growth rings 0.3-3.0 mm broad;
tracheids in 1-9 radial rows between the narrow rays; early wood
tracheids individually of rather large diameter (35-70 μm),
polygonal or squared but with pronounced tendency toward rounding
on the "corners" (thus often with slight intercellular spaces
between), showing abrupt transition to late wood; late wood zone

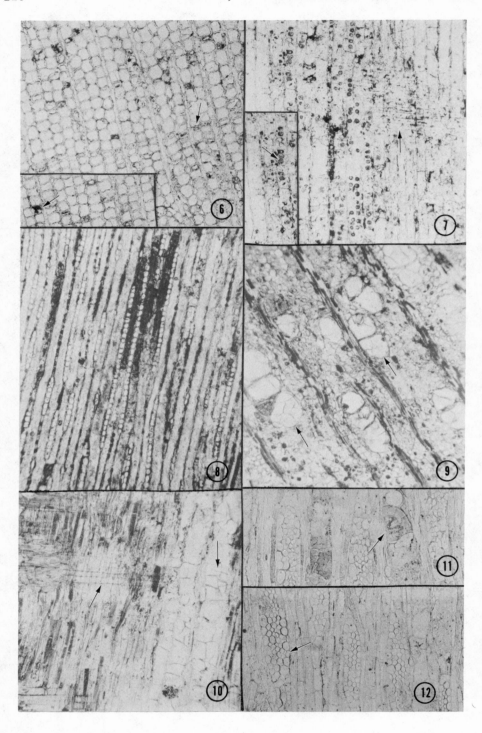

Fig. 6: Transverse section, Cupressinoxylon florense; arrow
pointing to slight growth ring; inset at lower left, arrow
pointing to resinous axial parenchyma cell.

Fig. 7: Radial section of C. florense; arrow pointing toward
ray; inset at lower left, arrow pointing to circular bordered
pit on tracheid wall.

Fig. 8: Tangential section of C. florense; note "tall"
uniseriate rays with more or less rounded cells; note also
resinous axial parenchyma strands.

Fig. 9: Transverse section of Floroxylon variabile; arrows
pointing to vessel multiples, tyloses visible inside.

Fig. 10: Radial section of F. variabile; arrow at left
pointing to ray parenchyma; arrow at right pointing to tyloses
inside a vessel.

Fig. 11: Tangential section of F. variabile; arrow pointing
toward "spherulite" of chalcedony inside a vessel.

Fig. 12: Tangential section of F. variabile; arrow pointing
to heterogeneous ray transected.

shallow (often 2-5 cell layers), the tracheids flattened yet tending to maintain their "rounded" shape on the corners; resin ducts (even traumatic ones) totally lacking; resin cells (axial parenchyma) scattered throughout the growth increments, often obvious by virtue of their dark contents.

Radial Section (Fig. 7): Intertracheal pitting uniseriate or occasionally biseriate-opposite in part; pits large (to 25 µm in diameter), circular-bordered, with a smooth torus, somewhat spaced vertically, often separated by Sanio's rims; wood parenchyma strands resinous, the cells with smooth (not obviously pitted) transverse walls; ray parenchyma smooth-walled, lacking indentures, showing cupressoid cross-field pitting to the axial tracheids; ray tracheids not evident; resin ducts absent.

Tangential Section (Fig. 8): Rays uniseriate, or sometimes biseriate in part, 2-45 cells high (thus some considered "tall"), the ray parenchyma cells individually rounded in cross-section; tangential intertracheal pitting sporadic, these pits smaller than those on the radial walls of the tracheids; resin cell chains (axial parenchyma strands) evident; resin ducts lacking.

Type = Blackwell: Collection B (MU).

Rapp (1954) correctly assigned this petrifaction to the fossil-wood genus Cupressinoxylon Goeppert (spelled "Cupressoxylon" in Rapp's thesis), even though not observing the critical para-meter of cross-field pitting (cf. Phillips, 1948; Jane, 1970). Similar species of Cupressinoxylon, differing only in minor details (e.g., ray height), have been described by Knowlton (1891), Berry (1922), Dukes (1961a) and others throughout the Southeast. None, though, could be viewed as truly conspecific with C. florense.

There is little difficulty in aligning the wood of Cupressinoxylon florense with the extant family Cupressaceae. The scattered axial parenchyma (resin cells) contribute to this determination and would be considered abundant by the criterion of Phillips (1948), i.e., five or more per square millimeter of cross-section surface. Peirce (1937) pointed out that a minority group within the Cupressaceae (Actinostrobus, Callitris, Tetraclininis, Callitropsis, and Widdringtonia) has wood parenchyma cells with smooth transverse walls, i.e., not showing the sculptured (beaded or nodular) thickenings generally characteristic of the family (cf. Juniperus). In having smooth-walled parenchyma, Cupressinoxylon florense corresponds more closely to this minority group in the family than to the more typical members. A similar observation was noted for C. nabortonensis by Dukes (1961a) from the Naborton and Dolet Hills Formation of Louisiana.

Since Cupressinoxylon (viz. C. florense) is considered (in our paper) identifiable to the Cupressaceae, i.e., assignable to a family, the name is used in a generic sense, rather than strictly than of a form-genus (fide Stafleu et al., 1978).

EUPHORBIACEAE:

FLOROXYLON variabile Rapp ex Blackwell in Blackwell et al.,
    gen. & sp. nov. (generic and specific description combined
    below; see Rapp, 1954, for source of name Floroxylon
    variabile)

Transverse Section (Fig. 9): Growth rings not or faintly evident, diffuse porous; vessels solitary or more typically in radial multiples of 2-5(-7) cells, in a given multiple the terminal vessel elements often the broadest, the successive radial multiples tending in turn to show radial alignment with each other; vessel elements ± oval, or angular by compression, the diameters varying from 50-150 µm; tyloses abundant, readily apparent, non-sclerotic; wood parenchyma very scanty, para-tracheal; rays numerous, multiseriate (to 6 cells wide) and uniseriate; wood fibers abundant and medium-walled (wall 4-6 µm thick).

Radial Section (Fig. 10): Vessel perforation plates simple, the end-wall inclination low; intervascular pitting alternate, crowded, the pits 9-10 µm in diameter, the pit borders hexagonal; tyloses evident in the vessel lumina; wood fibers somewhat storied, occasionally septate, averaging ca. one-half mm in length; rays sometimes with dark contents, procumbent and upright cells readily apparent.

Tangential Section (Figs. 11 and 12): Rays decidedly hetero-cellular (heterogeneous type I and II, see Kribs, 1935), weak, asymmetric, somewhat storied, with very elongate (possibly formerly lactiferous) upright marginal cells (stringers) often extending vertically for considerable distances, sometimes connecting vertically with rays above or below; wood fibers "flexible" and following a wavy axial course between the rays; vessel lines sinuous.

Type = Blackwell: Collection D (MU).

There is nothing to preclude adoption of Rapp's (1954) proposed name for the Flora dicotyledonous petrified wood. However, Rapp did not assign Floroxylon to a particular family, but rather suggested the possibilities of relationship with several families, including the Rhamnaceae and the Lauraceae as likely candidates. Berry (1922) considered somewhat similar wood from the Eocene of Louisiana to have lauraceous affinities, viz. Laurinoxylon. However, the Flora angiosperm petrifaction would very doubtfully belong to either the Rhamnaceae or the Lauraceae. The Rhamnaceae, for example, have very small-diametered vessels lacking tyloses (Record, 1939), and the Lauraceae could be expected to show a difference in vessel pattern, in intervessel pitting, and in a number of other features such as the usual presence of axial oil cells (Record and Hess, 1942). The Flora wood corresponds well in general terms with the overall description by Record (1938) for American Euphorbiaceae, and for that matter with the recently described wood of the euphorbiaceous genus Picrodendron (Hayden, 1977). Examination of members of the Euphorbiaceae in the wood slide collection in the Jodrell Laboratory at Kew (Richmond-Surrey, England) by one of us (Blackwell) has bolstered our belief in the euphorbiaceous character of the fossil dicot wood from Flora.

The Flora angiosperm wood is distinguished from the rather well known fossil genus Heveoxylon (Euphorbiaceae), cf. Tidwell (1975), by rays often more than biseriate and by the diffuse-porous "increments". Floroxylon actually bears a closer similarity to Paraphyllanthoxylon, assigned tentatively to the Lauraceae (as "Lauraceae?") in Tidwell's book. One might question the actual affinities of Paraphyllanthoxylon.

By describing Floroxylon herein and assigning it to the family Euphorbiaceae, the generic name is not only validated nomenclaturally, but also is altered from the status of a form-genus to that of a fossil genus assignable to a family (see International Code of Botanical Nomenclature, 1978).

Mineral Determinations

In examining thin-sections of the angiosperm material from Flora, radiating fibrous structures were sometimes observed, particularly within the vessels. These "spherulites" are especially apparent in radial section, although they are typically evident in other types of sections as well (Fig. 11). Under polarized light, they are anisotropic against an isotropic background. With more difficulty, they can also be observed in the gymnosperm petrifaction. Macroscopically, the mineral in question is somewhat chalky or waxy, often, with proper lighting, giving evidence of very small crystals. In fungal decayed cavities in the "wood", a grape-like (botryoidal) crystal growth cluster pattern can be observed. These evidences indicate the replacement mineral to be a form of micro- or crypto-crystalline quartz known as chalcedony (cf. Kerr, 1977; Chesterman, 1978). This is in fact the mineral found in the "trees" in the famous petrified forest in Arizona (Chesterman, 1978). However, a more

or less isotropic (scarcely birefringent) background material is
also usually present. This has been determined to be opal, and
is as well commonly found in fossil woods from various parts of
the United States (Chesterman, 1978). It is not surprising that
chalcedony and opal would occur together as replacement minerals
since both are low temperature silicates (SiO_2) differing
primarily in the degree of hydration, opal being hydrous silicon
dioxide. Thus, the chief replacement minerals in the Flora woods
appear to be mixtures of chalcedony and opal. Kerr (1977)
indicated that the two minerals are commonly associated. In
the angiospermous petrifaction chalcedony seems more abundant
than in the coniferous petrifaction, which apparently is predominantly opalized. The vessel diameters in the angiosperm
would probably permit greater growth of the spherulites of
chalcedony, than would the diameters of the tracheids of the conifer.

Individual cell walls (of, for example, tracheids or fibers)
examined from either petrifaction under higher magnifications
with a polarizing microscope appear isotropic, i.e., primarily
composed of opal. There is no hint of a cellulose crystalline
structural differential as might be observed in the S_1, S_2, and
S_3 layers of secondary wall respectively (Jane, 1970) due to the
"pitch" of cellulose microfibrils. The uniformity of the "cell
walls" of the petrifactions under polarized light supports Dukes'
belief (1961a) that no cellulose or true wall material is left.
It should be noted, however, that degraded lignin compounds were
identified in the Triassic age silicified Araucarioxylon trunks
from the Petrified Forest National Park in Arizona (Sigleo, 1978).
Thus, further chemical experimentation with the Mississippi
petrified logs might be in order.

Phytogeographic Implications

This study has turned up no information which would support
the idea that the silicified logs at Flora were of "boreal"
origin, i.e., having arrived at their present location as a
result of a long river journey from "the North". The two fossil
woods described herein match well with the two photomicrographs
published in Priddy's (1960) survey; however, they are not
respectively Abies and Acer (as indicated in Priddy's work),
and would seemingly have nothing to do with an Arcto-Tertiary
Geoflora, as fir and maple most certainly did (Vankat, 1979). By
contrast, both Flora woods seem to show tropical characteristics
and to have affinities which are primarily tropical. There are
a number of evidences for this including, for example, in the
angiosperm material, the absence or indefiniteness of growth
rings, the rather highly storied wood with relatively weak (sometimes septate) wood fibers, and the stringer-like vertical ray

extensions. The botanical relationship of the angiospermous
petrifaction with arborescent Euphorbiaceae would in itself indicate a tropical correlation.

Tropical or subtropical relationships for the Flora woods
should not really be surprising in that the climate of the times,
especially as far south as the latitude of Central Mississippi,
generally did not have the temperature minima extremes which occur
at the present time. As MacGinite (1953) stated in connection
with work on the Florissant Beds in Colorado, "all evidence points
to higher average temperatures in western or southern states in
pre-Pliocene times than obtain today". That warm, moist climatic
conditions were the rule in the South can be attested to by the
abundance of Tertiary fossil palm material at several different
localities in Louisiana and Mississippi (e.g., Knowlton, 1888;
Blackwell, personal observation).

The dicotyledonous (euphorbiaceous) petrifaction would
definitely appear to have Neotropical relationships. Its
closest extant counterparts are perhaps to be sought in northern
South America and/or the West Indies. For example, wood slides
of Alchorneopsis (Euphorbiaceae) examined at Kew bore a fairly
close similarity to this wood. There are two species of
Alchorneopsis, A. portoricensis in the West Indies and A.
floribunda in the Amazon basin. Although we are not suggesting
that Alchorneopsis is the closest living relative of Floroxylon,
and basically hold to the belief that such assessments are often
unwise if not unwarranted, a genus like Alchorneopsis possesses
a type of wood and a pattern of distribution which would "fill
the bill" of relationship better at least than previous suggestions.

The coniferous (cupressoid) petrifaction may actually tell
something about continental drift, or at least a former continental connection between the southeastern United States and
northwestern Africa. That such a connection existed is probable
(Hurley, 1968). The extant genera of Cupressaceae mentioned to
which Cupressinoxylon florense shows the greatest similarity,
i.e., those having wood parenchyma with smooth end-walls, do not
occur in the New World today but rather are found from Northwest Africa (Tetraclinis) to South Africa (Widdringtonia) to
West Australia (Actinostrobus) to New Caledonia (Callitropsis),
cf. Peirce (1937) and Phillips (1948). It is interesting that
these smooth-walled genera are considered primitive within the
family Cupressaceae (Gregus, 1955). Wood specimens of
Tetraclinis examined at Kew and photographs of the wood of
Callitropsis (Gregus, 1955) do show a close similarity to the
Flora coniferous material, differing really only in minor details.

The fossil genus Cupressinoxylon extends back to Triassic time
and was quite widespread (Seward, 1919; Blackwell, personal
observation of a variety of slides at the Paleontological Research
Section of the British Museum of Natural History, London). Thus,
it is possible that an ancestral type of cupressoid tree existed
in a once-joined area between the southeastern United States and
northwestern Africa. By this hypothesis, today relic descendents
would persist, for example, in Northwest Africa and New Caledonia,
but only fossil remains would be found in the southeastern United
States. The severe climatic change that occurred through late
Tertiary time in North America would have accounted for the
extinction of forms here. Cupressinoxylon florense thus may be
another at least minor botanical evidence for continental drift.

We view the logs at Flora as probably having been once part
of a tropical (perhaps swamp) forest type which may well have
existed at, or close to, the subsequent site of fossilization.
This opinion should not be unrealistic since, as discussed,
similar types of "wood" occur throughout the south (especially
western Mississippi, Louisiana, Arkansas and East Texas).
Material of Cupressinoxylon described by Knowlton (1891) from
Arkansas, for example, was definitely "growing" in situ, and did
not arrive at the site as driftwood of distant origin. The
Flora petrifactions would seem to have no obvious relationship
with Arcto- or Madro-Tertiary species found at the time much
farther to the north and/or west.

It is of interest to note that in addition to the locality at
Flora, Mississippi, at least one other occurrence of the Forest
Hill Formation has been studied botanically (Berry, 1924). This
was an outcrop located at "Rocky Hill Church", about 10 miles
north of Jackson, Mississippi, and therefore not too distant
from Flora. It is not known whether or not this site still
exists, or could be "unearthed". Regardless, Berry reported a
list of plants from it based apparently on leaf impressions and
at least one florule. It is probably labored to repeat the list
herein; however, the taxa included in it (e.g., Ficus unionensis)
would appear to have predominantly tropical connections. This
would, of course, correspond with our general belief concerning
the phytogeographic correlations of the material at Flora. We
believe that the Forest Hill Sand, in its various outcrops,
deserves further attention in the future. Study of this and
similar formations might well lead to a reevaluation of concep-
tions of the vegetation of the southeastern United States
during Tertiary time.

ACKNOWLEDGMENTS

We wish to thank Dr. David M. Scotford of the Department of Geology, Miami University, for his help with the mineral determinations discussed in this paper, and Mr. Joe H. Marak (Curator, Geology Museum) for his assistance with the use of the thin-section equipment. We also thank Dr. Martha J. Powell of the Department of Botany for her identification of the fossil fungus mentioned, and Dr. Charles Heimsch for his aid in interpretation of wood structure. We are especially grateful to the Schabilion family of Mississippi, owners of the Petrified Forest, for their interst, hospitality, and encouragement throughout the course of this investigation.

LITERATURE CITED

Ainsworth, B. D. 1967. Minerological and grain-size data on selected samples from the Forest Hill formation in western Mississippi. U. S. Army Corps of Engineers (Vicksburg, MS) Misc. Paper No. 6-916.

Berry, E. W. in D. White. 1922. Additions to the flora of the Wilcox group. U. S. Geol. Sur., Prof. Paper 131: 1-22.

Berry, E. W. 1924. The middle and upper Eocene floras of south-eastern North America. U. S. Geol. Sur., Prof. Paper 92: 1-206.

Brown, C. S. 1913. The petrified forest of Mississippi. Popular Sci. Monthly 83: 466-470.

Chesterman, C. W. 1978. The Audubon society field guide to North American rocks and minerals. Alfred A. Knopf, New York.

Dukes, G. H. 1959. Some Pleistocene fossil woods of central Mississippi. Unpublished master's thesis, Mississippi College.

Dukes, G. H. 1961a. Some Tertiary fossil woods of Louisiana and Mississippi. Dissertation, Louisiana State University.

Dukes, G. H. 1961b. Some Tertiary fossil woods of Louisiana and Mississippi. Amer. J. Bot. 48: 540. (abstract)

Federal Writers' Project. 1938. Mississippi, a guide to the magnolia state. American Guide Series, Viking Press, New York. (republished in 1973 by Somerset Publishers, St. Clair Shores, Michigan)

Goeppert, H. R. 1850. Monographie der fossilen coniferen. Naturwerkundige Verhand. Maatschap. Wettenschappen Haarlem, Leiden.

Greguss, P. 1955. Identification of living gymnosperms on the basis of xylotomy. Akademiai Kiado, Budapest.

Hayden, W. J. 1977. Comparative anatomy and systematics of Picrodendron, genus incertae sedis. J. Arnold Arb. 58(3): 257-279.

Hildebrand, J. R. 1937. Machines come to Mississippi. Nat. Geogr. Mag. 72(3): 263-318.

Hurley, P. M. 1968. The confirmation of continental drift. Sci. Amer. 218(4): 52-64.

Jane, F. W. 1970. The structure of wood. Adam and Charles Black, London. (second edition revised by K. Wilson and D. J. B. White)

Kerr, P. F. 1977. Optical minerology, fourth edition. McGraw-Hill Book Co., New York.

Knowlton, F. H. 1888. Description of two species of Palmoxylon -- one new -- from Louisiana. Proc. U. S. Nat. Mus. 11: 89-91.

Knowlton, F. H. 1891. Cupressinoxylon calli. Geol. Sur. Ann. Rept. Ark., 1889, Vol. 2: 254, pl. 9, figs. 3-7.

Kribs, D. A. 1935. Salient lines of structural specialization in the wood rays of dicotyledons. Bot. Gaz. 96: 547-557.

MacGinitie, H. D. 1953. Fossil plants of the Florissant beds, Colorado. Carnegie Inst. Washington, Publ. 599.

Peirce, A. S. 1937. Systematic anatomy of the woods of the Cupressaceae. Trop. Woods 49: 5-21.

Penhallow, D. P. 1907. Notes on fossil woods from Texas. Royal Soc. Can., Trans., 4: 93-113.

Phillips, E. W. J. 1948. Identification of softwoods by their microscopic structure. For. Prod. Res. Bull. 22: 1-56. (reprinted 1970, Her Majesty's Stationery Office, London)

Priddy, R. R. 1960. Madison county geology. Miss. State Geol. Sur. Bull. 88: 1-123.

Rapp, L. H. 1954. Some petrified woods of Mississippi. Unpublished master's thesis, Univ. of Cincinnati.

Record, S. J. 1938. The American woods of the family Euphorbiaceae. Trop. Woods 54: 7-40.

Record, S. J. 1939. American woods of the family Rhamnaceae. Trop. Woods 58: 6-24.

Record, S. J. and Hess, R. 1942. American timbers of the family Lauraceae. Trop. Woods 69: 7-33.

Schabilion, S. 1973. We bought a petrified forest. Rock & Gem 3(10): 20-24.

Seward, A. C. 1919. Fossil Plants 4: 1-543. Cambridge University Press. (Hafner Reprint, 1963)

Shaw, E. W. 1918. The Pliocene history of northern and central Mississippi. U. S. Geol. Sur., Prof. Paper 108: 125-163.

Sigleo, A. C. 1978. Degraded lignin in silicified wood 200 million years old. Science 200: 1054-1055.

Stafleu, F. A. et al. (editors). 1978. International code of botanical nomenclature (adopted by the Twelfth International Botanical Congress, Leningrad, July 1975). Regnum Vegetabile Vol. 97. Bohn, Scheltema and Holkema, Utrecht.

Suzzallo, H. (ed.). 1932. The national encyclopedia. Vol. 8. p. 23. P. F. Collier and Son Co., New York. (reprinted, 1949).

Tidwell, W. D. 1975. Common fossil plants of western North America. Brigham Young University Press, Provo, Utah.

Vankat, J. L. 1979. The natural vegetation of North America, an introduction. John Wiley and Sons, New York.

Wilmarth, M. G. 1938. Lexicon of geologic names of the United States. USDI Bull. 896, Part 1, A-L.

CUPULE ORGANIZATION IN EARLY SEED PLANTS

Lawrence C. Matten* and William S. Lacey[+]

*Dept. of Botany, Southern Illinois Univ.
 Carbondale, Illinois
[+]School of Plant Biology
 Univ. College of North Wales
 Bangor, Wales

ABSTRACT

    Multiovulate cupules from the Upper Devonian and Lower
Carboniferous have been studied to determine the morphology of the
cupules. Species studied include Archaeosperma arnoldii,
Calathospermum fimbriatum, C. scoticum, Eurystoma angulare,
Geminitheca scotica, Gnetopsis elliptica, Hydrasperma longii,
H. tenuis, and Stamnostoma huttonense. The cupule of
Archaeosperma arnoldii, the earliest seed, is interpreted as
bearing four ovules (rather than two) and being composed of oval
to terete (rather than flattened) units producing approximately
16 distal segments. The cupules can be characterized by the
branching pattern of the cupule segments. Some species
(Archaeosperma arnoldii, Stamnostoma huttonense,) are characterized
by successive dichotomies of the cupule stalk forming the enve-
loping cupule. Other species (Calathospermum fimbriatum,
Eurystoma angulare, Hydrasperma tenuis) are characterized by an
initial dichotomy of the cupule stalk followed by monopodial
(pseudomonopodial?) divisions (pinnate) forming the cupule.
Geminitheca scotica is characterized by a combination of the two
types of patterns. The cupule units of Calathospermum scoticum,
Gnetopsis elliptica, and Hydrasperma longii are fused at their
base. In addition, the cupules are compared by size, position
of stalk, and number of included ovules (seeds).

221

Seeds (ovules) first occur in sediments of the Upper Devonian and Lower Carboniferous. The oldest seeds, Archaeosperma arnoldii and Hydrasperma tenuis, are cupulate indicating the early appearance, and primitiveness of the cupule. The possible evolutionary importance of the cupule and its apparent co-evolution with the seed suggests the need for a comparative morphological study.

The occurrence of a large collection of ovulate cupules of Hydrasperma tenuis (Matten, Lacey, Edwards, 1975; Matten, Lacey, Lucas, 1978, 1980; Matten, Lacey, May, Lucas, 1980) has permitted the opportunity to observe such characteristics as variation in branching morphology and preservation. This study has now been extended to include a comparison of the branching patterns with other ovulate cupules, in particular those Paleozoic forms bearing two or more ovules per cupule. The species included in this study are: Archaeosperma arnoldii (Pettitt and Beck, 1968), Calathospermum scoticum (Walton, 1949), Calathospermum fimbriatum (Barnard, 1960), Geminitheca scotica (Smith, 1959), Eurystoma angulare (Long, 1960b, 1965), Hydrasperma tenuis (Long, 1961; Matten, Lacey, Edwards, 1975; Matten, Lacey, Lucas, 1980), Hydrasperma longii (Long, 1977a, 1977b, 1979; Matten, Lacey, Lucas, 1980), Gnetopsis elliptica (Renault, 1885), and Stamnostoma huttonense (Long, 1960a).

Archaeosperma arnoldii (Pettitt and Beck, 1968) is, to date, the oldest seed to be described. The specimens are preserved as compressions and the reconstruction of Pettitt and Beck shows a pair of cupules on a common stalk with the cupule units being somewhat flattened. An alternative interpretation is that the specimen figured by Pettitt and Beck represents a single cupule and that the cupule units are more rounded (Figs. 1, 2). Compressions of Archaeosperma show what appear to be flattened cupule units, especially at the base of the cupule. A similar appearance was observed in compressions of Hydrasperma tenuis cupules but peel sections showed that the units were rounded and only subsequently flattened, producing a preservation artifact. It is thought that the cupule units of Archaeosperma underwent the same process as those of Hydrasperma tenuis and that the present reconstruction represents a more accurate reflection of the cupule's morphology. Identification of terete, oval, or flattened structures from compressions adds one more problem to interpretations by paleobotanists. Thus alternate reconstructions must be considered. In the case of terete versus flattened cupule units in the Upper Devonian Archaeosperma, terete units seem to be a more logical choice than flattened.

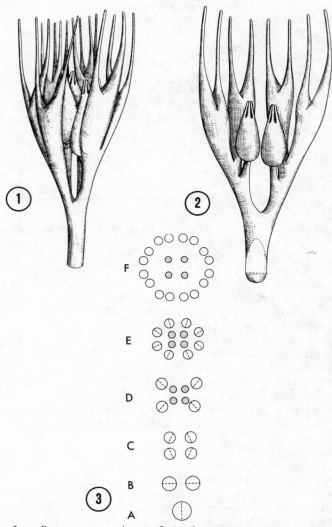

Figure 1. Reconstruction of Archaeosperma arnoldii as a
four-seeded cupule. Figure 2. Inner view of half a cupule of
Archaeosperma arnoldii. Figure 3. A series of diagramatic cross
sections (A-F) through the Archaeosperma arnoldii cupule. The
drawings are based on the description of Pettitt and Beck (1968)
and show a dichotomous arrangement of cupule units.

A series of diagramatic cross sections through the
Archaeosperma cupule, based on the description of Pettitt and
Beck (1968) shows a dichotomous system (Fig. 3). The erect
cupule stalk forks successively approximately four times producing up to 16 cupular units at the distal end. The cupule of
Archaeosperma is about 15 mm long and 7-8 mm wide. Each cupule
contains four seeds (ovules).

The cupule of Stamnostoma huttonense (Figs. 7,8), has been
interpreted as developing from successive dichotomies of the
cupule axis forming as many as 16 sterile units at the distal end.
The cupules are about 14 mm long and 4 mm wide. They are erect
on their stalk and are thought to bear four seeds.

The branching patterns of the cupules of Eurystoma,
Hydrasperma, Calathospermum and Geminitheca seem to vary from
the basic dichotomous patterns observed in Archaeosperma and
Stamnostoma. Barnard's (1960) interpretation of the morphology
of the cupule of Calathospermum fimbriatum seems to be the most
acceptable framework with which to compare these ovulate
cupules. Barnard visualizes the cupule of Calathospermum
fimbriatum (Fig. 6) as representing a terminal portion of a
frond. The cupule is divided into two halves which may correspond to the two main arms (secondary rachises) typical of
many pteriodsperm fronds. Barnard equates the base of the cupule
to the region of bifurcation (Barnard, 1960). Each of the
secondary rachises gives rise to a pair (pairs) of lateral
primary pinnae and ends in a terminal pinna. The primary
pinnae form the basic segments of the cupule. Long (1965)
illustrates a series of transverse sections through a pair of
Calathospermum fimbriatum cupules that supports Barnard's interpretation. The cupules of C. fimbriatum are up to 90 mm long and
up to 30 mm wide. There may be as many as 64 free cupule units
at the distal end and the cupule may contain as many as 24 seeds.

The cupules of Hydrasperma tenuis (Matten et al., 1975;
Matten et al., 1980) are arranged in pairs on a common stalk.
Each cupule is erect, and contains 2-6 seeds (Figs. 9,10,12).
The cupules are 5.4-10.4 mm long and 7.1-11.0 mm wide. Serial
transverse sections through four-seeded and two-seeded cupules
show similar division of the cupule stalk into cupule units
(Figs. 11,13,14). The cupule stalk first bifurcates (equivalent
to Barnard's bifurcation of the pteridosperm frond). Each of the
resultant cupular units (secondary rachises) then produces a
subopposite pair of lateral pinnae (in sectional view it looks
like a trifurcation or two very close dichotomies). The result

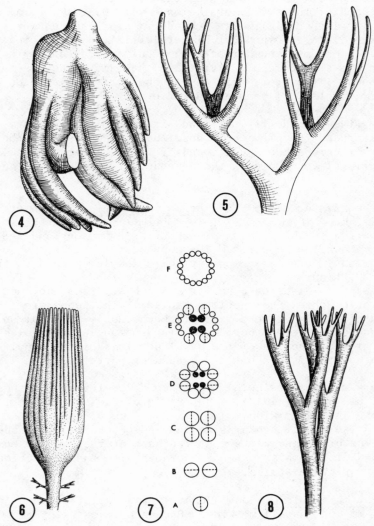

Figure 4. Reconstruction of Hydrasperma longii cupule
(redrawn from Long, 1977a). Figure 5. Reconstruction of
Eurystoma angulare cupule. The seeds have been removed for
viewing the arrangement of the cupule units (redrawn from Long,
1965, cupule A). Figure 6. Reconstruction of Calathospermum
fimbriatum (redrawn from Barnard, 1960). Figure 7. A series of
diagramatic cross sections (A-F) through the Stamnostoma
huttonense cupule showing the arrangement of cupule units (redrawn from Long, 1960a). Figure 8. Reconstruction of cupule of
Stamnostoma huttonense. The seeds have been removed for ease of
viewing the cupule units (redrawn from Long, 1960a).

is a total of six cupule units at this level. In the four-seeded
cupule, each of the two lateral pinnae will bear a lateral fertile
pinnule (stalked ovule). The secondary rachis subsequently produces other lateral pinnae at higher levels (in an alternate
pattern). The total number of cupule units produced may exceed
18. The two-seeded cupule divides in the same manner as the four-
seeded cupule. However, only one of the lateral pinnae on each
secondary rachis will bear a stalked ovule (Fig. 13).

The five-seeded (Fig. 15) and six-seeded cupules show the
same pattern as the previously described cupules except that the
secondary rachises bear additional "fertile" lateral pinnules at
a higher level (one more in the 5-seeded, two more in the 6-seeded
cupules).

A second species of Hydrasperma, H. longii (Matten et al.,
1980), consists of incurved cupules having coherent basal
units (Fig. 4). The cupules are about 9-10 mm long and 5-8 mm
wide and are borne in pairs. The cupule stalk has a v-shaped
xylem strand in cross-sectional view (the xylem strand in the
stalk of H. tenuis is terete). The cupule is composed of two
basally united halves each having 6-8 lobes. Thus the cupule
has up to 16 lobes (one specimen had 18). The lobes are flattened basally and terete distally. Each cupule may bear up to 16
ovules and one specimen bore both microsporangia and ovules.
Because of the cohesion of the basal portions of the cupule not
much can be discerned about its branching pattern.

The same comments hold for a second species of Calathospermum,
C. scoticum (Walton, 1949). The cupule of C. scoticum is large,
45 mm long and 24 mm wide, with six major cupule units fused
basally. The cupule contains as many as 48 ovules. Although
there is no branching pattern at the base of the cupule segments
there is a dichotomous system associated with the production of
the numerous ovule stalks.

Eurystoma angulare (Long 1960b, 1965) consists of cupules
about 15 mm long and 11 mm wide. Each cupule bears up to 10
ovules. The number of cupule units may be as high as 15. The
cupule units may be slightly flattened near their base but are
generally terete in the distal region. Examination of three reconstructed cupules of Eurystoma(Long, 1965) called specimens A,
B, and C, shows that the branching pattern of each cupule is
similar to that of Hydrasperma tenuis.

The pattern of Eurystoma angulare specimen A (Fig. 5) is essentially that of a two-seeded Hydrasperma tenuis cupule. In
summary, the cupule axis first dichotomizes (forming two secondary

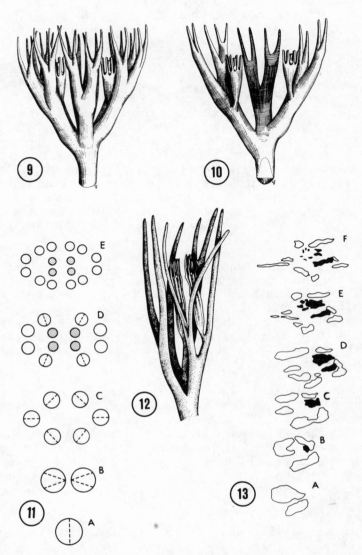

Figure 9.   Reconstruction of four-seeded cupule of
Hydrasperma tenuis.   Figure 10.   Inner view of half a cupule of
Hydrasperma tenuis.   Figure 11.   A series of diagramatic cross
sections (A-E) through the Hydrasperma tenuis four-seeded
cupule.   Figure 12.   Reconstruction of a two-seeded cupule of
Hydrasperma tenuis.   Figure 13.   A series of cross sections
(A-F) through a two-seeded cupule of Hydrasperma tenuis showing
the outline of the cupule units and seeds (black areas).

rachises) and then each secondary rachis produces lateral
pinnae. One lateral pinna of each secondary rachis bears a
sessile seed. The other lateral pinnae and the secondary rachis
may bear sterile laterals. A total of 11 sterile cupule units are
present in specimen A.

The cupule of Eurystoma angulare specimen B (Figs. 16,17)
is more complex than that of specimen A but still shows the
branching pattern of the Hydrasperma tenuis cupule. The sequence
of divisions follows the pattern of the six-seeded cupule of H.
tenuis. A major difference is the vertical distance between
divisions of each of the two secondary rachises. After the initial
bifurcation of the stalk, each secondary rachis bears at least
three fertile lateral pinnae. In addition the production of additional vegetative lateral pinnae and pinnules results in perhaps
as many as 20 cupule units. The delayed divisions of one half of
the cupule may be due to the somewhat recurved nature of the
cupule on its stalk.

Long's (1965) specimen C of Eurystoma angulare is interpreted as being composed of two cupules. Cupule $S_1S_2S_3$ (Fig. 18)
divides in the pattern previously described for specimens A and B.
One half of the cupule bears four seeds and is composed of at
least six sterile units. The other half of the cupule bears two
seeds and is composed of at least three sterile units. Cupule
$S_4S_5S_6$ (Fig. 19) shows a similar but less developed branching
pattern as that for the other cupules.

Geminitheca scotica (Smith, 1959) represents both microspor-
angiate and ovulate cupules. Only the latter are considered here.
They are borne in pairs. Each cupule base divides into two
secondary rachises which then divide in either a dichotomous or
monopodial (pseudomonopodial?) pattern. A series of transverse
sections through a pair of cupules (Fig. 20) shows the left one
to follow the pattern of Hydrasperma tenuis and results in six
cupule units. The cupules are 9-12 mm long and 2-4 mm wide and
bear two seeds.

Gnetopsis elliptica (Renault, 1885) has a cupule that is
comparatively quite small, 6 mm long and 4 mm wide. Each cupule
bears at least four seeds. The base of the cupule is united but
divided into two major parts each of which may have five or six
lobes. The stalk of the cupule lies approximately at right
angles to the central axis of the cupule (Walton, 1949). The
cupule of Gnetopsis elliptica is thought to be similar to that
of Calathospermum scoticum (Walton, 1949).

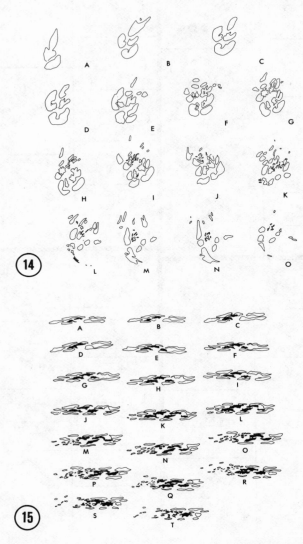

Figure 14. A series of cross section (A-O) through a four-seeded cupule of Hydrasperma tenuis. The four seeds (4,5,6,7) can be seen in section F. The division of the cupule stalk (A) is followed by each half forming 3 of the 6 main units of the cupule (B and E). Figure 15. A series of cross sections (A-T) of a five-seeded cupule of Hydrasperma tenuis. The seeds are in black.

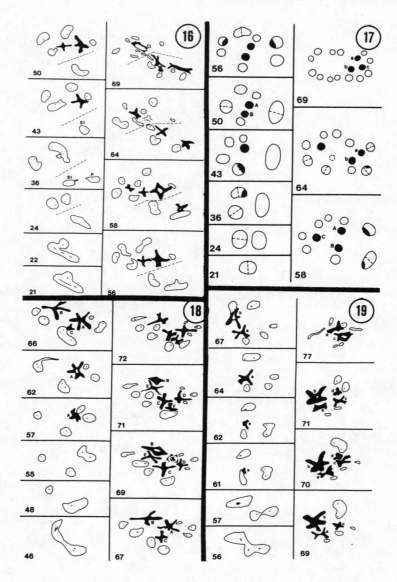

Figure 16. A series of cross sections through a six-seeded cupule of <u>Eurystoma angulare</u>. This is specimen B of Long (1965) and the numbers refer to his original numbering system (redrawn from Long, 1965, textfigures 3 and 4). Figure 17. Interpretive drawings of sections of <u>Eurystoma angulare</u> (cupule B) illustrated in Fig. 16. Figure 18. A series of cross sections through specimen C (cupule $S_1S_2S_3$) of <u>Eurystoma angulare</u> (redrawn from Long, 1965, textfigures 6-8). The numbering system corresponds to that of Long's textfigures. Figure 19. A series of cross

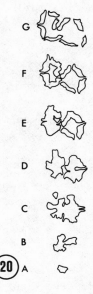

Figure 20.   A series of cross sections through a pair of
cupules of <u>Geminitheca</u> <u>scotica</u> (redrawn from Smith, 1959).

sections through specimen C (cupule $S_4S_5S_6$) of <u>Eurystoma</u> <u>angulare</u>
(redrawn from Long, 1965, textfigures 6-8).   The numbering system
corresponds to that of Long's textfigures.

Within these few pages all of the petrified, multiovulate cupules have been enumerated. In addition to these eight species, Archaeosperma arnoldii has been added because of its significance. These nine species have several characteristics in common. The cupules generally occur in pairs, perhaps even in clusters of pairs (e.g., Stamnostoma). The cupules are stalked and thought to occur at the ends of branch systems. Each cupule unit is vascularized and, except for Calathospermum scoticum, each cupule has a dichotomy at its base (i.e., top of stalk).

Variation among the species includes number of ovules per cupule (2-48), number of cupule units (6-64), position of cupule on its stalk (erect to bent at right angles to recurved), size 6 x 4 mm to 45 x 24 mm, attachment of ovules (sessile in Eurystoma to stalked in Calathospermum), xylem shape in cupule stalk, (v-shaped in Hydrasperma longii and terete in H. tenuis), and seed morphology.

The branching patterns of the cupule units can be used to divide the species into groups. The group that has a dichotomous pattern above the first bifurcation includes Archaeosperma arnoldii, Stamnostoma huttonense, and Geminitheca scotica. The group that has a monopodial (pseudomonopodial?) pattern above the basal dichotomy of the cupule stalk includes Eurystoma angulare, Hydrasperma tenuis and Calathospermum fimbriatum (perhaps also Geminitheca scotica in part). A third (and admittedly artificial) group would include those species whose cupule units were fused at their base, Calathospermum scoticum, Gnetopsis elliptica and Hydrasperma longii.

If, as is generally proposed, the cupule units represent coalesced portions of a branch system then the types of variation seen in the cupules of these nine genera would indeed be pre-dictable. Non-fused, terete cupule units would be most like the primordial branching system and flattened and fused units would be a derived combination.

It is also interesting to note that the primitive cupules were multi-ovulate while the derived and younger forms are uni-ovulate.

ACKNOWLEDGEMENTS

We wish to express our appreciation to the following: Office of Research Development and Administration, SIU-C Graduate School for support of field work; Ms. Karen Schmitt for preparing the illustrations and reconstructions that appear in this paper; and Mr. John Richardson, Director, Scientific Photography and Illustration Facility, SIU-C for help with photography.

LITERATURE CITED

Barnard, P. D. W. 1960. Calathospermum fimbriatum sp. nov., a Lower Carboniferous pteridosperm cupule from Scotland. Palaeontology, 3: 265-275.

Long, A. G. 1960a. Stamnostoma huttonense gen. et. sp. nov. --- a pteridosperm seed and cupule from the Calciferous Sandstone Series of Berwickshire. Trans. Roy. Soc. Edin., 64: 201-215.

-------. 1960b. On the structure of "Samaropsis scotica" Calder (emended) and Eurystoma angulare gen. et. sp. nov., petrified seeds from the Calciferous Sandstone Series of Berwickshire. Trans. Roy. Soc. Edin., 64: 261-280.

-------. 1961. Some pteridosperm seeds from the Calciferous Sandstone Series of Berwickshire. Trans. Roy. Soc. Edin., 64: 410-419.

-------. 1965. On the cupule structure of Eurystoma angulare. Trans. Roy. Soc. Edin., 66: 111-128.

-------. 1975. Further observations on some Lower Carboniferous seeds and cupules. Trans. Roy. Soc. Edin., 69: 267-293.

-------. 1977a. Some Lower Carboniferous pteridosperm cupules bearing ovules and microsporangia. Trans. Roy. Soc. Edin., 70: 1-11.

-------. 1977b. Lower Carboniferous pteridosperm cupules and the origin of angiosperms. Trans. Roy. Soc. Edin., 70: 13-35.

-------. 1979. The resemblance between the Lower Carboniferous cupules of Hydrasperma cf. tenuis Long and Sphenopteris bifida Lindley and Hutton. Trans. Roy. Soc. Edin., 70: 129-137.

Matten, L. C., W. S. Lacey, and D. Edwards. 1975. Discovery of one of the oldest gymnosperm floras containing cupulate seeds. Phytologia, 32: 299-303.

Matten, L. C., W. S. Lacey and R. C. Lucas. 1978. Cupulate
    seeds of Hydrasperma from Kerry Head, Ireland. Bot. Soc.
    Amer. Misc. Publ. 156: 3.

-------. 1980. Studies on the cupulate seed genus Hydrasperma
    Long from Berwickshire and East Lothian in Scotland and
    County Kerry in Ireland. Bot. Jour. Linn. Soc. Lond. 81:
    249-273.

Matten, L. C., W. S. Lacey, B. I. May and R. C. Lucas. 1980. A
    megafossil flora from the Uppermost Devonian near Ballyheigue,
    Co. Kerry, Ireland. Rev. Palaeobot. Palynol. 29: 241-251.

Pettitt, J. M. and C. B. Beck. 1968. Archaeosperma arnoldii--
    a cupulate seed from the Upper Devonian of North America.
    Contr. Mus. Paleontol., Univ. Mich., 22: 139-154.

Renault, B. 1885. Genre Gnetopsis. Cours Bot. Foss., 4: 179-184.

Smith, D. L. 1959. Geminitheca scotica gen. et sp. nov.: a
    pteridosperm from the Lower Carboniferous of Dunbartonshire.
    Ann. Bot., N. S., 23: 477-491.

Walton, J. 1949. Calathospermum scoticum--An ovuliferous
    fructification of Lower Carboniferous age from Dunbartonshire.
    Trans. Roy. Soc. Edin., 61: 719-728.

PLANT SUCCESSION ON A FILLED SALT-MARSH AT CAPE ANN,

MASSACHUSETTS, 1958-1979

Ralph W. Dexter

Dept. of Biological Sciences, Kent State University

Kent, OH 44242

I.   Introduction

During September-November, 1958 the U.S. Army Corps of Engineers dredged Lobster Cove at Cape Ann, Massachusetts, when 17 acres of bottom sediments were removed and dumped onto a salt-marsh directly across the Annisquam Tidal River. A total of 184,120 cubic yards of sediment was spread over the marsh, and a V-shaped dike was constructed around the margin. The dike was 4-5 feet high, while the enclosed marsh was covered to a depth of 1.5-2.5 feet. Each year since then, the area has been examined and photographed to trace the successional development as the sediments became revegetated.

Studies on the successional development of salt-marshes on the northeast coast of North America have been published by Transeau (1909, 1913), Davis (1910), McAtee (1939), Chapman (1940, 1964) and Redfield (1965, 1972). Succession studies on reclaimed salt-marshes have been published by Ganong (1903), Deane (1915, 1926) and Ellis (1931).

II.   Annual Progress in Succession, 1959-1979

Three levels were recognized. The lower level, which was subject to some inundation during periods of extreme spring tides (two openings were left in the dike); the middle level consisting of hummocks and the margin above the level of the extreme spring tides; the higher level consisting of the dike

and the margin of the fill which joined a sandy beach. Only the
dominant vegetation was studied.[1]

1959 -- The first summer a few widely scattered pioneers
invaded the lowest level such as Salicornia europaea (glasswort),
Salsola kali (saltwort), Sueda linearis (sea blite), Cerastium
arvense (chickweed), and the short form of Spartina alterniflora
(thatch grass). A few domestic vegetables (tomato, cucumber,
squash, etc.) sprouted from garbage contained in the sediments.
Also, a few scattered plants of Ammophila breviligulata (beach
grass) and Solidago sempervirens (seaside goldenrod) grew on the
middle level. A few seedlings of beach grass and seaside
goldenrod also appeared on the dike.

1960 -- Hudsonia tomentosa (beach heather) invaded all
levels with increasing density of the above species.

1961 -- The short-form of S. alterniflora was spread widely
and S. patens (salt-marsh hay) appeared at the lower level.
Valiela et al. (1978) have shown that the growth form of S.
alterniflora is determined by available nutrients. H. tomentosa
was much more abundant and widely distributed.

1962 -- S. patens well established, but many bare areas
remained. Solidago and Ammophila more abundant generally, and
the dike was largely covered with Ammophila and Solidago.

1963 -- The lower level was essentially a S. patens marsh
with Atriplex patula (orach), Oenothera sp. (evening primrose),
Rumex crispus, R. pallida (dock), and with Salicornia persisting
in the bare areas. Agropyron pungens (sea-couch) formed a stand
in the corner of the dike. Myrica pensylvanica (bayberry) was
invading the hummocks at the middle level. On the dike Rhus
typhinia (staghorn sumac) began to invade; while at the highest
level where the fill joined a sandy beach, Populus deltoides
(cottonwood), Robinia pseudacacia (black locust), and Salix sp.
(willow) appeared at the margin.

1964 -- Phragmites communis (reed grass) was well established on the lower level. [Hitchcock (1833) did not list this
plant for Massachusetts. Robinson (1880) knew of only one small
stand of Phragmites in Essex County which he interpreted as an
introduced plant in the vicinity of Topsfield. In the course of

---

[1]Acknowledgment is made to the following for assistance with
identification of the plants: Elliott Rogers, John Kieran,
Dr. Tom Cooperrider, and Allison Cusick.

Fig. 1. South dike with pioneer invaders. Fill on the left. Aug., 1959.

Fig. 2.   General view of fill with pioneer invaders.   Sept., 1960.

Fig. 3.   Phragmites communis invading Spartina patens marsh in formation.   July, 1965.

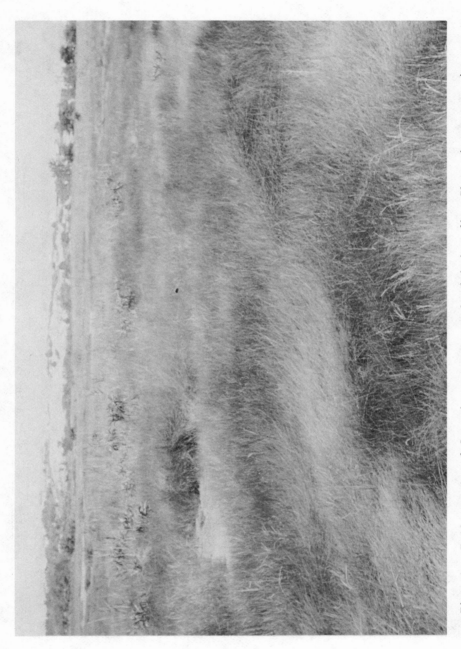

Fig. 4. Fully restored Spartina patens marsh with invading Phragmites communis. July, 1967.

my original study (1933-1937) on Cape Ann (Dexter, 1947) this
species was never encountered. Upon my return to re-sample the
marine life at Cape Ann in 1956, P. communis was abundant and
widely spread throughout the marshes of the Annisquam Tidal
Inlet.]. R. Typhinia was well established on the dike.

1965 -- Myrica pensylvanica (bayberry) well established, and
Lythium salicaria (purple loosestrife) and Lathyrus maritimus
(beach pea) appeared in the middle zone. Phragmites continued
expanding over the lower marsh.

1966 -- Opening in the dike enlarged. Entering spring tides
formed a ditch lined with short-form of S. alterniflora,
Phragmites invaded bayberry zone, which has replaced beach grass
on the middle level.

1967 -- With re-entry of salt water, the original S. patens
marsh was largely restored on lowest level. [For climax vegetation on marshes of the Annisquam Inlet see Dexter, 1947].
Around the periphery, Phragmites has formed a large stand.

1968 -- Four levels were well marked: S. patens, salt-
marsh; bayberry on the hummocks; beach grass on the dike;
cottonwood - black locust at the highest level.

1969 -- Phragmites invaded beach grass and bayberry on
middle level. Beach grass remained dominant on most of the
dike, and S. patens remained on low ground subjected to flooding
by extreme tides. Very few bare areas remain.

1970 -- Phragmites invaded beach grass on the dike. Lythrum
salicaria (purple loosestrife) and Agropyron pungens (sea-couch)
invaded edge of marsh. Cottonwood invaded staghorn sumac at
highest level.

1971 -- Terrestrial vegetation expanding at the expense of
salt-marsh grasses.

1972 -- Agrostis scabra (tickle grass) and Solidago
tenuifolia (goldenrod) appeared in the middle level, and Pinus
resinosa (red pine) and Juniper virginiana (red cedar) were
growing at the highest level.

1973 -- Phragmites now covered about 50% of filled marsh.

1974 -- Phragmites covered nearly 75% of filled marsh and
invaded the dike where sumac is still dominant.

Fig. 5. East dike with Rhus typhinia invading Ammophila breviligulata zone.
         Sept., 1974.

Fig. 6. Phragmites obliterating S. patens marsh. Sept., 1975.

Fig. 7. Solid stand of Phragmites encroaching on south dike. June, 1976.

Fig. 8. Four zones at east end. 1) Restricted S. patens marsh, 2) Myrica pensylvanica on hummocks, 3) Solid stand of Phragmites, 4) Populus deltoides and Robinia pseudoacacia on highest level. June, 1977.

Fig. 9. _Populus_ advancing into _Phragmites_ over western end of fill. July, 1979.

1975 -- S. patens found only in a small area flooded by spring tides. Phragmites continuing to spread everywhere.

1976 -- Northwest end of filled marsh a solid stand of Phragmites, behind which is a stand of cottonwood.

1977 -- Remaining stand of S. patens being invaded by Phragmites.

1978 -- Bare areas have disappeared.

1979 -- During 21 years three levels with four zones have developed: (1) S. patens marsh now largely obliterated by Phragmites; (2) beach grass largely replaced on hummocks by bayberry, and by sumac and Phragmites on the dike; (3) bayberry patches being replaced by Phragmites; (4) a stand of cottonwood and black locust slowly advancing at the landward margin.

LITERATURE CITED

Chapman, V. J. 1940. Succession on the New England salt marshes. Ecol. 21: 279-282.

--------. 1964. Coastal Vegetation. 245 pp. (Oxford Press).

Davis, C. A. 1910. Salt marsh formation near Boston and its significance. Econ. Geol. 5: 623-639.

Deane, Walter. 1915. Floral changes in a salt marsh during reclamation. Rhodora 17: 205-222.

--------. 1926. Further changes in a salt marsh during re-clamation. Rhodora 28: 37-40.

Dexter, R. W. 1947. The marine communities of a tidal inlet at Cape Ann, Mass.: a study in bio-ecology. Ecol. Monog. 17: 261-294.

Ellis, A. E. 1931. A reclaimed salt-marsh. Proceed. Malacol. Soc. London 19: 278-279.

Ganong, W. F. 1903. The vegetation of the Bay of Fundy salt and diked marshes: an ecological study. Bot. Gaz. 36: 161-186; 280-302; 349-367; 429-455.

Hitchcock, Edward. 1833. Report on the Geology, Minerology, Botany and Zoology of Mass. 700 pp. (Amherst, Mass.).

McAtee, W. L.   1935.   Wildlife of the Atlantic Coast salt
      marshes.   U.S.D.A., Bur. Biol. Survey, Wildlife Research
      Leaflet BS-17.   28 pp.

Redfield, A. C.   1965.   The ontogeny of a salt marsh estuary.
      Science 147: 50-55.

-------.   1972.   Development of a New England salt marsh.   Ecol.
      Monog. 42: 201-237.

Robinson, John.   1880.   The Flora of Essex County, Mass.   200 pp.
      (Essex Institute, Salem).

Transeau, E. N.   1909.   Successional relations of the vegetation
      about Yarmouth, Nova Scotia.   Plant World 12: 271-281.

-------.   1913.   The vegetation of Cold Spring Harbor, Long
      Island.   I.   The littoral successions.   Plant World 16:
      189-209.

Valiela, Ivan, J. M. Teal, and W. G. Deuser.   1978.   The nature
      of growth forms in the salt marsh Spartina alterniflora.
      Amer. Nat. 112: 461-470.

THE ENVIRONMENTAL DISTRIBUTION OF SOME LATE PRECAMBRIAN MICROBIAL ASSEMBLAGES

Andrew H. Knoll

Oberlin College

Oberlin, OH

ABSTRACT

Three sources of data are available for the paleoecological investigation of Precambrian microbes:

1. The spatial distribution and orientation of organisms in the rock (as seen in thin section);
2. the grouping of taxa in recurrent associations attributable to community patterns, transportation and mixing, and/or synonymy; and
3. the relationship of these associations to the sedimentary environment.

The application of these approaches casts discussions of Precambrian microbial diversity in a new light as it becomes possible to look at taxonomic diversity within a community, community distribution within an environment, and environmental heterogeneity on the ancient earth.

Several discrete benthonic associations of taxa can be recognized within flat, laminated, stromatolitic cherts from the Ross River locality of the 740-950 m.y. old Bitter Springs Formation, Australia. Additionally, a low diversity plankton assemblage is preserved among the mat microbes. Approximately coeval deposits from the Draken Conglomerate, Svalbard (collected by the Cambridge Spitsbergen Expedition) contain a similar series of benthonic associations, but a very different - and far more diverse - set of planktonic microfossils. Sedimentological

evidence suggests that periodic storms brought open shelf waters
into the Draken Lagoon, introducing planktonic algae and re-
freshing the lagoonal waters. New collections of samples from
the Svalbard Precambrian, which represent a wide variety of
depositional environments, broaden the documentation of the
ecological complexity of the late Precambrian earth and, when
applied to a range of time planes, can provide a proper context
for the evolutionary interpretation of the early fossil record.

CRETACEOUS POLLEN AND EARLY ANGIOSPERM EVOLUTION

James A. Doyle

University of California

Davis, CA

ABSTRACT

Analysis of the stratigraphic distribution and morphological
interrelationships of angiosperm pollen types from Cretaceous
sedimentary sequences of widely separated geographic areas provides a coherent framework for understanding of the timing,
course, and ecological context of early angiosperm evolution.
The recognizable angiosperm pollen record begins with the
appearance of monosulcate pollen with typically angiospermous
exine features (columellar infratectal structure and/or endexine
nonlaminated and restricted to the apertural area) in Barremian
rocks of both southern Laurasia and Africa-South America. The
ensuing diversification of monosulcates presumably reflects the
initial radiation of monocots and magnoliid dicots, although most
orders and families probably did not differentiate until much
later. Most reticulate monosulcate complexes occur in both
Laurasia and Gondwana, but apparently related tectate types,
showing transitions between columellar structure and the
putatively more primitive granular structure seen in some
magnoliids, are known only from Africa-South America. The first
tricolpate pollen types, from the Aptian of Africa, have
reticulate-columellar structure and show anomalies in aperture
arrangement suggesting derivation from earlier monosulcates.
Vigorous diversification of tricolpates in the Albian of both
sides of the Tethys results in a great diversity in size, shape,
and sculpture, and the first examples of the tricolporate
condition characteristic of most modern dicots. This phase
corresponds to the initial radiation of nonmagnoliid dicots:

relatively primitive tricolpate pollen is retained by some
Ranunculidae and "lower" Hamamelididae, while some late Albian
tricolporates could represent ancestral Rosidae and/or
Dilleniidae. Many modern "Amentiferae" may be derivatives of
the triporate Normapolles complex, which first appears in the
middle Cenomanian of Laurasia, probably derived from triangular
tricolporate ancestors; their smooth exines and granular
infratectal structure must be viewed as secondary specializations.
The observed radiation of angiosperm pollen types within the
Cretaceous implies that any pre-Cretaceous angiosperms must
have been very primitive and undiversified. The facies distribution of pollen types suggests that Early Cretaceous angiosperms were most abundant in certain lowland, but not coastal,
environments. The trend for reduction of exine sculpture in
the Normapolles and some other Cenomanian groups may represent
an adaptation for wind pollination, as inferred for recently
described Cenomanian inflorescences, but the well-developed
sculpture of earlier monosulcates and tricolpates supports the
concept that angiosperms were originally insect pollinated.

MID-HOLOCENE HEMLOCK DECLINE: EVIDENCE FOR A PATHOGEN OR INSECT

OUTBREAK

Margaret Bryan Davis

University of Minnesota

Minneapolis, MN

ABSTRACT

Hemlock (Tsuga canadensis) pollen percentages and influx
$cm^{-2}yr^{-1}$ show a sharp decline at all sites in eastern North
America 4800 ^{14}C yrs B.P. Hemlock pollen remained at low
levels for 1000 years and then started to increase, reaching its
original percentage abundance by 3000 yrs B.P. The hemlock
pollen decline is followed by a series of changes in pollen
abundances: first birch, then beech and maple pollen increase
in northern New Hampshire, birch then pine and oak in southern
New Hampshire, birch then pine in Michigan. These changes in
abundance correspond to the expected regional secondary
succession that would follow a unilateral reduction in the
hemlock population. The synchroneity of the hemlock decline
over a very large region, more than 10^6 km^2, and the successional
sequences that follow it suggest the spread of a pathogen that
reduced the hemlock population and kept it at low levels for
2000 years, until resistance developed or the virulence of the
pathogen was attenuated allowing the hemlock population to
increase again.